LIBRARY PUBLISHING DIRECTORY 2014

EDITED BY SARAH K. LIPPINCOTT

1230 PEACHTREE STREET, SUITE 1900
ATLANTA, GA 30309
WWW.LIBRARYPUBLISHING.ORG
919.533.9814
SARAH@EDUCOPIA.ORG

978-0-98991-180-1 (PRINT)
978-0-98991-181-8 (EPUB)
978-0-98991-182-5 (EPDF)

CONTENTS

LIBRARIES OUTSIDE THE UNITED STATES AND CANADA

FOREWORD

*Martin Halbert (University of North Texas),
James Mullins (Purdue University), and
Tyler Walters (Virginia Tech)*

In January 2013, we officially launched the Library Publishing Coalition (LPC) project, a collaborative initiative that now involves 60 academic libraries committed to advancing the emerging field of library publishing. As this new service area matures and expands, we have seen a clear need for knowledge sharing, collaboration, and development of common practices. The LPC is helping this field move forward in a number of key ways, but we are particularly proud to publish the first edition of the *Library Publishing Directory*, a guide to the publishing activities of 115 academic libraries.

In documenting the breadth and depth of activities in this field, this resource aims to articulate the unique value of library publishing; to establish it as a significant and growing community of practice; and to raise its visibility within a number of stakeholder communities, including administrators, funding agencies, other scholarly publishers, librarians, and content creators. Collecting this rich set of data from libraries across the United States and Canada allows us to identify themes, challenges, and trends; make predictions about future directions; and position the Library Publishing Coalition to better meet the needs of this community.

The *Directory* also advances one of the central goals of the Library Publishing Coalition, to facilitate and encourage collaboration among libraries as well as among libraries and publishers that share their values, especially university presses and learned societies. We hope that libraries will use the *Directory* to learn about their peers, find mutually beneficial ways to work together, and ultimately improve their practices and enhance the value they provide to their campuses. We hope that presses will see opportunities to initiate new partnerships or expand existing ones.

The LPC is a community-driven project that relies on the hard work and expertise of representatives from our participating institutions. We could not have produced the *Directory* without the support of the Directory Subcommittee. We would like to thank Marilyn Billings (University of Massachusetts-Amherst), Stephanie Davis-Kahl (Illinois Wesleyan University), Adrian Ho (University of Kentucky), Holly Mercer (University of Tennessee), Elizabeth Smart (Brigham Young University), Shan Sutton (Oregon State University), Allegra Swift (Claremont University Consortium), Beth Turtle (Kansas State University), and Charles Watkinson (Purdue University) for their invaluable contributions.

We also are grateful for the generous support of Purdue University Libraries' Scholarly Publishing Services unit, which donated resources, staff time, and the expertise of Alexandra Hoff and Managing Editor Katherine Purple; Lightning Source, who donated print-on-demand services; and the Charlesworth Group for conversion to ebook formats.

Finally, we would like to thank the libraries that took the time to help us to better understand, promote, and assert the significance of library publishing initiatives by providing information for this *Directory*. The 115 libraries listed in this first edition demonstrate the tremendous interest and energy in this field. We look forward to continuing to watch and document library publishing services as they evolve and progress in the coming years.

INTRODUCTION

Sarah K. Lippincott, Katherine Skinner, and Charles Watkinson

We are so pleased to share with our readership this first *Library Publishing Directory*, produced by the Library Publishing Coalition (LPC) in our organization's inaugural year of work. This *Directory* intends to make visible the innovation, support, and services offered today by a broad range of academic libraries in the area of scholarly communications.

Herein, we begin to document the strategic investments university libraries around the world are making in the area of academic publishing. Once believed to be a one-off activity subsidized by a small number of libraries, "Library Publishing" today is evolving into a dynamic subfield in the academic publishing ecosystem.

WHY PUBLISH A DIRECTORY?

For more than two decades, faculty, researchers, and students have come to their college and university libraries to gain technical support and staffing for early experiments in digital scholarship. From hosting ejournals and electronic theses and dissertations (ETDs) to collaborating with teams of researchers to construct multimedia experiences, these libraries have been willing and able partners in this academic mission of creating and disseminating scholarship.

By 2007, these library-based activities began to formalize, as documented in two key reports: Ithaka S&R's *University Publishing in a Digital Age* and ARL's *Research Library Publishing Services: New Options for University Publishing.* Subsequent studies reinforced the importance of these emerging library-based publishing endeavors. As demonstrated by the seminal *Library Publishing Services: Strategies for Success* report, publishing services now are thriving across the whole range of academic libraries today, from small liberal arts colleges to premier research institutions.

This growth of library publishing activities provided the impetus and rationale for creating the LPC to help advance this subfield for U.S. and Canadian academic libraries. Hosted by the Educopia Institute, and driven by 60 academic libraries, the LPC project (2013–2014) is now founding this new organization. Its mission is to promote the development of innovative, sustainable publishing services in academic and research libraries to support scholars as they create, advance, and disseminate knowledge. As a key part of this work, the LPC seeks to document practices and services in the field, and to foster strategic alliances and connections both across and between libraries and other academic publishers.

The LPC created this *Directory* to begin to answer the many questions the project team had about the publishing activities currently underway in libraries. How many libraries define their scholarly communications activities as "publishing"? How long have they been doing this work? With whom do they partner? What types of publications are they producing? Are libraries offering specific products and/or services to their campuses? What percentage of their publications are peer reviewed? How many staff members are working on this activity, and how are they funding their activities? Are there identifiable models and trends in this subfield of publishing today?

With these and other questions in mind, the LPC Directory Subcommittee built and disseminated an Internet-based survey in spring 2013, targeting North American listservs for academic libraries. We focused on North America for scoping reasons: we knew we could not hope to chronicle global work in full, and so began with this smaller-but-significant subset of activity. We intentionally structured this *Directory* to encompass institutions beyond the LPC itself, inviting any institution engaged in library publishing to participate. We received more than 110 responses to this survey. In the following pages, we include directory listings for all institutions that responded, grouping the North American institutions first (our primary target) and programs outside the U.S. and Canada next. Using the survey data, the LPC Directory Subcommittee assembled the directory entries, shared each one with its institutional representative for editing and approval, and then published it herein. We greatly appreciate all those who gave their time and energy to help us document the efforts of their individual libraries. Notably, the only institutions listed here are those that responded to our survey. Undoubtedly, many important programs have been missed in this first edition. We hope that those we have missed will contact LPC (sarah@educopia.org) so we can ensure these institutions are included in future editions.

The *Library Publishing Directory* contributes directly to the LPC's goal of encouraging collaboration by allowing library publishing staff, who have traditionally had relatively little contact with each other, to identify colleagues producing scholarly work in similar disciplines or using the same technology platform. The *Directory* also is intended to open the way to collaboration with other publishers, especially mission-driven non-profit university presses and learned societies, by introducing and articulating the unique and complementary approach that libraries take to the publishing function. Finally, it is hoped that the *Directory* can help scholarly authors to become more aware of the opportunities that may exist on their own campuses or in their disciplines to experiment with new publication formats or business models.

We highlight below some of the exciting library publishing trends and models we see emerging in this first *Directory* of activity. Together the answers provide a rich picture of what types of product libraries are creating and what technological, financial, and human resources they are using.

LIBRARY PUBLISHING TODAY

Individually, the *Directory* entries reveal much about local practices, including the mission driving an institution's activities, the funding models and staffing supporting its work, the relationship between publication and preservation, and the type and quantity of publications produced.

Collectively, these entries say far, far more. Last year, the libraries profiled in this *Directory* published 391 faculty-driven journals, 174 student-driven journals, 937 monographs, at least 8,746 conference papers and proceedings, and nearly 100,000 each of ETDs and technical/research reports. These publications covered an array of disciplines, including law, agriculture, history, education, computer science, and many, many others. Thirty-three libraries report disciplinary specialties in the social sciences and area, ethnic, cultural, and gender studies (a broad classification that includes a range of interdisciplinary specialties). Education (24 libraries), health and clinical sciences (22), and the general humanities (18) are also particularly well-represented areas.

Faculty-driven journals were the most common publication reported by these libraries. Over 70% of the libraries in this *Directory* published at least one in 2012 and over half (54%) published at least one student-driven journal. Thirty-six percent produced at least one monograph, and more than three-quarters published ETDs. More than half reported publishing data, audio, and video, in addition to text and images.

Currently, there is no single, dominant model for the organization of publishing services. In many institutions, services are distributed across multiple library units or across campus. The lead unit varies across libraries (e.g., Scholarly Communications, Technical Services, and even Special Collections). Library publishing programs featured in this *Directory* range from small, experimental endeavors to large, more mature operations with several dedicated staff members. Libraries reported between .01 and eight full-time equivalent in library staff, and many also reported employing graduate (30%) and undergraduate (26%) students.

Across these libraries, the most prominent services are building, implementing, maintaining, and supporting publishing platforms for authors. In this work many report using full-service digital platforms, including Public Knowledge Project's OJS/OCS/OMS suite (45%), bepress's Digital Commons platform (34%), and DSpace (34%)—the top three for respondents. However, many also report developing software locally (20%) and/or using a content management system like WordPress (18%) for dissemination and delivery. More than three-quarters of respondents said that they provide a broader range of services, including metadata (85%), analytics (62%), outreach (51%), DOI assignment (42%), audio/video streaming (42%), and ISSN registration (40%). A substantial number of these libraries also provide support for editorial and production processes.

These include peer review management (29%), copyediting (23%), and print-on-demand (16%). Finally, some libraries support business model development (13%), budget preparation (6%), and contract and license preparation (28%). Other services offered, such as author advisory on copyright (78%), build upon librarians' strengths as educators and advocates.

A hallmark of library publishing, as is repeatedly highlighted in individual directory entries, has been the building of partnerships with content creators and other publishers on and off campus. Faculty, students, and other authors typically provide the editorial leadership for library publications. Over 85% of libraries featured herein report that they have relationships with campus departments or programs; 93% partner with individual faculty; and over half work with graduate and undergraduate students. Many of the libraries in this *Directory* report that they work with or have administrative ties to university presses. Off-campus partners include scholarly societies, non-profit organizations, museums, library networks and consortia, and individual faculty at other institutions.

Despite the different forms library publishing activities have assumed to date, the *Directory* demonstrates that these programs share a growing commonality of philosophy and approach combining traditional library values and skills (such as a concern with long-term preservation, expertise in the organization of information, and commitment to widening access) with lightweight digital workflows to create a distinctive "field" of publishing activity. The libraries in this *Directory* overwhelmingly prefer open access publication (95% focus mostly or completely on open access). And although 90% of libraries rely in part or completely on their library's operating budget to support publishing services, notably, 10% do not. Among those libraries that are subsidizing these activities, the operating budget is contributing an average of 88% of the publishing budget.

Looking toward the future, many of the libraries in this *Directory* report that they plan to increase the numbers and types of publications they produce, support an expanded suite of services (particularly in areas like data management), identify new partners within and beyond campus, and make improvements to software and workflows.

THE FUTURE OF LIBRARY PUBLISHING

As libraries undertake the improvement and expansion of services, they will continue to confront a difficult and rapidly changing landscape. Building capacity, sustaining services, and securing funding will require concerted efforts to demonstrate value and improve business models. Raising credibility and visibility on campus and within the broader scholarly communications community will also require individual and collective efforts. Libraries will need to convince campus administrators, university presses, librarians, commercial publishers, and content creators that library publishing is an important, strategic, purposeful service area that adds value to the publishing ecosystem. Perhaps most important, libraries will need to cultivate and strengthen their relationships with other scholarly publishers—including university presses, scholarly societies, and commercial publishers—to build our collective capacity, extend the reach of scholarship, and ensure that the scholarly communication apparatus continues to evolve in pace with the research and knowledge produced across academia.

This 2014 *Library Publishing Directory* tells a compelling story, one that we believe needs dissemination in its own right. We look forward to seeing these networks continue to build upon the work they have done. We hope the *Directory* will help existing and prospective library publishers identify new partners and learn from the experiences of their colleagues. And, of course, we hope to see the nexus of activity represented here continue to expand in the years ahead.

LIBRARY PUBLISHING COALITION SUBCOMMITTEES

The following Subcommittee Members have donated their time and expertise to advancing the Library Publishing Coalition's mission and producing its most significant resources.

PROGRAM SUBCOMMITTEE

The Program Subcommittee bears primary responsibility for planning and implementing the Library Publishing Forum.

Sarah Beaubien (Grand Valley State University)
Dan Lee (University of Arizona)
Mark Newton (Columbia University)
Melanie Schlosser (Ohio State University)
Marcia Stockham (Kansas State University)
Allegra Swift (Claremont University Consortium)
Evviva Weinraub (Oregon State University)

DIRECTORY SUBCOMMITTEE

The Directory Subcommittee provides support for the design and creation of the Library Publishing Directory.

Marilyn Billings (University of Massachusetts-Amherst)
Stephanie Davis-Kahl (Illinois Wesleyan University)
Adrian Ho (University of Kentucky)
Holly Mercer (University of Tennessee)
Elizabeth Smart (Brigham Young University)
Shan Sutton (Oregon State University)
Allegra Swift (Claremont University Consortium)
Beth Turtle (Kansas State University)
Charles Watkinson (Purdue University)

RESEARCH SUBCOMMITTEE

The Research Subcommittee coordinates Library Publishing Coalition Roundtable Discussions, and manages the organization's research agenda

Donna Beck (Carnegie Mellon University)
Marilyn Billings (University of Massachusetts-Amherst)
Brad Eden (Valparaiso University)
Isaac Gilman (Pacific University)
Dan Lee (University of Arizona)
Gail McMillan (Virginia Tech)
Catherine Mitchell (California Digital Library)
Jane Morris (Boston College)
Melanie Schlosser (Ohio State University)
Mary Beth Thompson (University of Kentucky)

READING AN ENTRY: SOME "HEALTH WARNINGS"

The field of library publishing is rapidly evolving, and its boundaries have not yet been clearly defined. We have attempted to produce a directory that is readable and cohesive and that allows for cross-institutional comparison. In some cases, this means we have used terminology and categories that do not fully reflect the complex and experimental nature of activities that libraries are undertaking. We hope that through this *Directory*, and through input from the library publishing community, we will start to establish common language as this field matures.

In some cases, as described below, questions in the questionnaire on which the entries are based were not specific enough and respondents reported numbers in different ways. Revised questions will be sent in future years.

While "staff in support of publishing activities" are consistently reported for salaried employees, the way in which respondents reported students percentages may vary. For example, undergraduate employees usually work a maximum of 20 hours per week rather than the 40 expected of a full-time staff member. Therefore, it is open to question whether a 0.25 undergraduate works 5 or 10 hours per week.

Under "types of publication," we have noticed that conference proceedings have been reported in various ways. In some cases, respondents report the number of series, while in other cases, they report the total number of faculty papers. Some high estimates of total number of publications raise an issue of whether the instructions to just record activity in the last full calendar year, 2012, were followed.

Readers will notice the presence of "seals" next to the title of some entries. These acknowledge the support of the institutions that fund the Library Publishing Coalition by their generous two-year pledges. "Contributing Institutions" have pledged to support the foundation of the LPC with an annual contribution of $1,000. "Founding Institutions" receive the highest honor, having pledged $5,000 a year to the project. To recognize their exceptional contributions, we include profiles of specific publications that Founding Institutions have nominated. These also give a practical sense of the wide range of types of publications produced.

Contributing Institution — Library Publishing Coalition

Founding Institution — Library Publishing Coalition

LIBRARIES IN THE UNITED STATES AND CANADA

ARIZONA STATE UNIVERSITY

Hayden Library

Primary Unit: Informatics and Cyberinfrastructure Services

Primary Contact:
Mimmo Bonanni
Digital Projects Manager
480-965-8168
digitalrepository@asu.edu

Website: repository.asu.edu

Social media: @asulibraries; facebook.com/ASULibraries

PROGRAM OVERVIEW

Mission/description: Arizona State University Libraries created the ASU Digital Repository to support ASU's commitment to excellence, access, and impact. The ASU Digital Repository advances the New American University by providing a central place to collect, preserve, and discover the creative and scholarly output from ASU faculty, research partners, staff, and students. Providing free, online access to ASU scholarship benefits our local community, encourages transdisciplinary research, and engages scholars and researchers worldwide, increasing impact globally through the rapid dissemination of knowledge. The ASU Digital Repository improves the visibility of content by exposing it to commercial search engines such as Google, the ASU Libraries' One Search, as well as the ASU Digital Repository search portal. The ASU Digital Repository helps meet public access policies and archival requirements specified by many federal grants.

Year publishing activities began: 2011

Organization: centralized library publishing unit/department

Staff in support of publishing activities (FTE): library staff (4.5); undergraduate students (1)

Funding sources (%): library operating budget (100)

PUBLISHING ACTIVITIES

Types of publications: faculty-driven journals (1); technical/research reports (503); databases (2); ETDs (1242); undergraduate capstone/honors theses (449)

Media formats: text; images; audio; video; data; concept maps/modeling maps/visualizations

Disciplinary specialties: state and local documents (government publications); music; dance

Top publications: *Journal of Surrealism* (journal)

Campus partners: campus departments or programs; individual faculty

Publishing platform(s): CONTENTdm; locally developed software

Digital preservation strategy: digital preservation services under discussion

Additional services: outreach; metadata; DOI assignment/allocation of identifiers; author copyright advisory; digitization; audio/video streaming

Plans for expansion/future directions: Improving marketing and outreach (involving Subject Librarians and E-Research staff), expanding data management support, and exploring the addition of learning objects.

AUBURN UNIVERSITY

Auburn University Libraries

Primary Contact:
Aaron Trehub
Assistant Dean for Technology and Technical Services
334-844-1716
trehuaj@auburn.edu

PROGRAM OVERVIEW

Mission/description: To support the university's outreach mission by making original research and scholarship by Auburn University faculty and students more accessible to Alabama residents and the world at large.

Year publishing activities began: 2008

Organization: services are distributed across campus

Staff in support of publishing activities (FTE): library staff (3)

Funding sources (%): library operating budget (90); charitable contributions/ Friends of the Library organizations (10)

PUBLISHING ACTIVITIES

Types of publications: ETDs (512)

Media formats: text; images

Top publications: ETDs

Campus partners: campus departments or programs; individual faculty

Publishing platform(s): DSpace

Digital preservation strategy: digital preservation services under discussion. Auburn University Libraries is a founding member of two Private LOCKSS Networks (MetaArchive Cooperative; ADPNet), but does not currently use these distributed digital preservation networks to preserve ETDs or materials in the IR.

Additional services: graphic design (print or web); outreach; training; cataloging; metadata; open URL support; author copyright advisory; digitization; audio/video streaming

Plans for expansion/future directions: Currently building and populating an institutional repository.

BOSTON COLLEGE

Boston College University Libraries

Library Publishing Coalition — Contributing Institution

Primary Unit: Scholarly Communications

Primary Contact:
Jane Morris
Head of Scholarly Communications and Research
617-552-4481
jane.morris@bc.edu

Website: www.bc.edu/libraries/collections/eScholarshipHome

PROGRAM OVERVIEW

Mission/description: Our goal is to showcase and preserve Boston College's scholarly output and to maximize research visibility and influence. eScholarship@BC encourages community contributors to archive and disseminate scholarly work, peer-reviewed publications, books, chapters, conference proceedings, and small datasets in an online open access environment.

Year publishing activities began: 2006

Organization: services are distributed across library units/departments

Staff in support of publishing activities (FTE): library staff (2.5)

Funding sources (%): library operating budget (100)

PUBLISHING ACTIVITIES

Types of publications: faculty-driven journals (1); student-driven journals (2); journals produced under contract/MOU for external groups (5); faculty conference papers and proceedings (1); ETDs (170); undergraduate capstone/honors theses (50); grey literature from centers/institutes; datasets

Media formats: text; video; data

Disciplinary specialties: theology; education; the Middle East; libraries

Top publications: *Catholic Education* (journal); *Studies in Christian-Jewish Relations* (journal); *Information Technology and Libraries* (journal); *Levantine Review* (journal); *Proceedings of the Catholic Theological Society of America* (conference proceedings)

Percentage of journals that are peer reviewed: 50

Campus partners: campus departments or programs; individual faculty; graduate students; undergraduate students

Other partners: Catholic Theological Society of America; ALA Library and Information Technology Association; Council of Centers on Christian Jewish Relations; Seminar on Jesuit Spirituality

Publishing platform(s): OJS/OCS/OMP; DigiTool

Digital preservation strategy: HathiTrust; LOCKSS; MetaArchive

Additional services: marketing; training; analytics; cataloging; metadata; ISSN registration; DOI assignment/allocation of identifiers; dataset management; contract/license preparation; author copyright advisory; digitization; audio/video streaming

Plans for expansion/future directions: Hosting more data and developing more open access journals.

BRIGHAM YOUNG UNIVERSITY

Harold B. Lee Library

Primary Unit: Scholarly Communication Unit
scholarsarchive@byu.edu

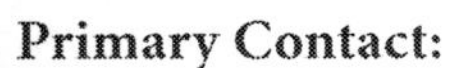

Primary Contact:
Elizabeth Smart
Scholarly Communication Librarian
801-422-4995
elizabeth_smart@byu.edu

Website: sites.lib.byu.edu/scholarsarchive

PROGRAM OVERVIEW

Mission/description: The Harold B. Lee Library's primary publishing resources include an institutional repository and digital publishing services for faculty- and student-edited journals. Combined, these resources are called ScholarsArchive. ScholarsArchive is designed to make original scholarly and creative work—such as research, publications, journals, and data—freely and persistently available. The library's publishing efforts are targeted at supporting broader academic and public discovery and use of university scholarship. ScholarsArchive may also house items of historic interest to the university. The library supports content partners with software support, digitizing, metadata creation, journal management, and free hosting services.

Year publishing activities began: 2001

Organization: centralized library publishing unit/department

Staff in support of publishing activities (FTE): library staff (2); undergraduate students (0.5)

Funding sources (%): library operating budget (98); charge backs (2)

PUBLISHING ACTIVITIES

Types of publications: faculty-driven journals (9); student-driven journals (3); journals produced under contract/MOU for external groups (3); student conference papers and proceedings (1); databases (1); ETDs (640)

Media formats: text; images

Disciplinary specialties: religion; natural history of the American West; children's literature

Top publications: *Western North American Naturalist* (journal); *BYU Studies* (journal); *Children's Book and Play Review* (journal); *Pacific Studies* (journal); *TESL Reporter* (journal)

Percentage of journals that are peer reviewed: 90

Campus partners: campus departments or programs; individual faculty; graduate students; undergraduate students

Other partners: International Society for the Comparative Study of Civilizations (ISCSC); Association of Mormon Counselors and Psychotherapists (AMCAP); Council on East Asian Libraries (CEAL)

Publishing platform(s): CONTENTdm; OJS/OCS/OMP

Digital preservation strategy: Rosetta (moving from beta to full implementation in 2013)

Additional services: analytics; cataloging; metadata; peer review management; digitization; hosting of supplemental content

Plans for expansion/future directions: Areas of future exploration and possible expansion include monograph publishing, print on demand, DOI support, hosting streaming media, and data management.

HIGHLIGHTED PUBLICATION

The *Western North American Naturalist* (formerly *Great Basin Naturalist*) has published peer-reviewed experimental and descriptive research pertaining to the biological natural history of western North America for more than 70 years.

ojs.lib.byu.edu/spc/index.php/wnan

BROCK UNIVERSITY

James A. Gibson Library

Primary Contact:
Elizabeth Yates
Liaison / Scholarly Communication Librarian
905-688-5550 ext. 4469
eyates@brocku.ca

Website: www.brocku.ca/library/about-us-lib/openaccess

PROGRAM OVERVIEW

Mission/description: The library's publishing initiatives provide technology, expertise, and promotional support for researchers, students, and staff at Brock University seeking to make their research universally accessible via open access. The library currently publishes/hosts five scholarly OA journals in partnership with Scholars Portal and the Ontario Council of University Libraries. We use Open Journal Systems (OJS) software. The library manages an open access publishing fund to help Brock authors cover the costs of publishing with fully OA journals or monograph publishers. A minimum of four awards of up to $2,500 are granted; total funding is $10,000. The library also hosts and disseminates Brock scholarship through our Digital Repository, which collects graduate theses, major research projects, and subject- or department-based research collections and materials from our Special Collections and Archives. We also raise awareness of open access through Open Access Week activities, information resources, and other venues.

Year publishing activities began: 2011

Organization: services are distributed across library units/departments

Staff in support of publishing activities (FTE): library staff (1)

PUBLISHING ACTIVITIES

Types of publications: faculty-driven journals (5); technical/research reports (3); ETDs (1887)

Media formats: text; images

Disciplinary specialties: humanities; French language; arts education; teaching and learning

Percentage of journals that are peer reviewed: 100

Campus partners: campus departments or programs; individual faculty

Other partners: Ontario Council of University Libraries/Scholars Portal

Publishing platform(s): DSpace; OJS/OCS/OMP

Digital preservation strategy: Scholars Portal

Additional services: copy-editing; training; analytics; notification of A&I sources; ISSN registration; digitization

Plans for expansion/future directions: Launching a journal showcasing undergraduate student research in the Faculty of Applied Health Sciences; launching an Open Monograph publishing system in partnership with Scholars Portal and the Ontario Council of University Libraries.

CAL POLY, SAN LUIS OBISPO

Robert E. Kennedy Library

Contributing Institution — Library Publishing Coalition

Primary Unit: Digital Scholarship Services

Primary Contact:
Marisa Ramírez
Digital Scholarship Services Librarian
805-756-7040
mramir14@calpoly.edu

Website: digitalcommons.calpoly.edu/; lib.calpoly.edu/scholarship

PROGRAM OVERVIEW

Mission/description: The Robert E. Kennedy Library provides digital services to assist the campus community with the creation, publication, sharing, and preservation of research, scholarship, and campus history.

Year publishing activities began: 2008

Organization: centralized library publishing unit/department

Staff in support of publishing activities (FTE): library staff (2); undergraduate students (3)

Funding sources (%): library operating budget (98); endowment income (2)

PUBLISHING ACTIVITIES

Types of publications: faculty-driven journals (6); student-driven journals (1); monographs (52); technical/research reports (102); faculty conference papers and proceedings (210); student conference papers and proceedings (160); newsletters (15); databases (1); ETDs (975); undergraduate capstone/honors theses (1841); graduate internship reports

Media formats: text; images; audio; video; data; multimedia/interactive content

Disciplinary specialties: history; philosophy; sustainability

Top publications: senior undergraduate projects; master's theses; *Between the Species* (journal); *California Climate Action Planning* (conference proceedings)

Percentage of journals that are peer reviewed: 100

Campus partners: individual faculty; undergraduate students

Publishing platform(s): bepress (Digital Commons)

Digital preservation strategy: digital preservation services under discussion. We are in the process of joining LOCKSS and MetaArchive.

Additional services: typesetting; copy-editing; marketing; outreach; training; analytics; cataloging; metadata; notification of A&I sources; ISSN registration; peer review management; business model development; author copyright advisory; other author advisory; digitization; hosting of supplemental content; audio/video streaming

Plans for expansion/future directions: Hiring an Endowed Digital Scholarship Services Student Assistant through the Digital Scholarship Services Student Assistantship Program, which provides paid, experiential learning opportunities for Cal Poly students who are interested the various facets of the changing digital publishing landscape.

CALIFORNIA INSTITUTE OF TECHNOLOGY

Caltech Library

Primary Unit: Metadata Services Group

Primary Contact:
Kathy Johnson
Repository Librarian
626-395-6065
kjohnson@library.caltech.edu

PROGRAM OVERVIEW

Mission/description: CaltechTHESIS is part of CODA, the Caltech Collection of Open Digital Archives, managed by Caltech Library Services. The mission of CODA is to collect, manage, preserve, and provide global access over time to the scholarly output of the Institute and the publications of campus units. CaltechTHESIS contains PhD, engineer's, master's, and bachelor's/senior theses authored by Caltech students. Most items in CaltechTHESIS are textual dissertations, but some may also contain software programs, maps, videos, etc. The ETD is the version of record for the Institute and deposit of doctoral dissertations is required for graduation.

Year publishing activities began: 2001

Organization: services are distributed across library units/departments

Staff in support of publishing activities (FTE): library staff (1)

Funding sources (%): library operating budget (100)

PUBLISHING ACTIVITIES

Media formats: text; images; audio; video; data; simple websites

Disciplinary specialties: biology; chemistry and chemical engineering; engineering and applied science; geology and planetary science; physics; mathematics; astronomy

Campus partners: campus departments or programs; individual faculty; graduate students; undergraduate students

Publishing platform(s): EPrints

Digital preservation strategy: digital preservation services under discussion

Additional services: marketing; outreach; training; analytics; cataloging; metadata; author copyright advisory; digitization

Plans for expansion/future directions: Undergoing gradual move of platforms to Islandora/Fedora, including preservation activity.

CALIFORNIA STATE UNIVERSITY SAN MARCOS

Kellogg Library

Primary Contact:
Carmen Mitchell
Institutional Repository Librarian
760-750-8358
cmitchell@csusm.edu

Website: csusm-dspace.calstate.edu; scholarworks.csusm.edu

Social media: @csusm_library

PROGRAM OVERVIEW

Mission/description: The purpose of the California State University San Marcos institutional repository (ScholarWorks) is to collect, organize, preserve, and disseminate CSUSM research, creative works, and other academic content in a web-based environment.

Year publishing activities began: 2011

Organization: centralized library publishing unit/department

Staff in support of publishing activities (FTE): library staff (1.4); undergraduate students (0.5)

Funding sources (%): library operating budget (95); other (5)

PUBLISHING ACTIVITIES

Types of publications: ETDs (250); library exhibits

Media formats: text; images; audio; video

Disciplinary specialties: student work/research; library exhibits

Top publications: "Going Paperless: Student and Parent Perceptions of iPads in the Classroom" (thesis); "Lateral Violence in Nursing" (thesis); "Nurses' Technique and Site Selection in Subcutaneous Insulin Injection" (thesis); "Individual Differences in Working Memory and Levels of Processing" (thesis); "Wounded Hearts: A Journey Through Grief" (thesis)

Campus partners: campus departments or programs; individual faculty; graduate students; undergraduate students

Publishing platform(s): DSpace

Digital preservation strategy: in-house; digital preservation services under discussion

Additional services: marketing; outreach; training; analytics; cataloging; metadata; contract/license preparation; author copyright advisory; other author advisory; digitization; hosting of supplemental content; audio/video streaming

Plans for expansion/future directions: Planning to include faculty publications and datasets within the next year, working with other CSU campuses on an undergraduate journal, and currently working to publish digital surrogates of items from the university archives.

CARNEGIE MELLON UNIVERISTY

Carnegie Mellon University Libraries

Primary Unit: Archives and Digital Library Initiatives

Primary Contact:
Gabrielle Michalek
Head of Archives and Digital Library Initiatives
412-268-7268
gabrielle@cmu.edu

Website: repository.cmu.edu

PROGRAM OVERVIEW

Mission/description: Carnegie Mellon University Libraries' publishing program aims to promote open access to scholarly resources, to support online journals and conference management—from article submission through peer review to open access and long-term preservation, and to publish grey literature, including theses, dissertations, and technical reports.

Year publishing activities began: 2010

Organization: services are distributed across library units/departments

Staff in support of publishing activities (FTE): library staff (1.5)

Funding sources (%): library operating budget (100)

PUBLISHING ACTIVITIES

Types of publications: faculty-driven journals (1); technical/research reports (2362); faculty conference papers and proceedings (1); ETDs (1988); undergraduate capstone/honors theses (79)

Media formats: text

Disciplinary specialties: social and behavioral sciences; engineering; physical and life sciences; arts and humanities; security

Top publications: *Journal of Privacy and Confidentiality* (journal); Dietrich College honors theses

Percentage of journals that are peer reviewed: 100

Campus partners: campus departments or programs; individual faculty; graduate students; undergraduate students

Publishing platform(s): bepress (Digital Commons)

Digital preservation strategy: LOCKSS; MetaArchive

Additional services: marketing; outreach; training; analytics; cataloging; metadata; peer review management; author copyright advisory; other author advisory; digitization; hosting of supplemental content; audio/video streaming

Plans for expansion/future directions: Hosting more open access journals; supporting conference proceedings; and publishing more theses, dissertations, and technical reports.

CLAREMONT UNIVERSITY CONSORTIUM

Claremont Colleges Library

Primary Unit: Center for Digital Initiatives
scholarship@cuc.claremont.edu

Primary Contact:
Allegra Swift
Digital Initiatives Librarian
909-607-0893
allegra_swift@cuc.claremont.edu

Website: scholarship.claremont.edu; ccdl.libraries.claremont.edu

Social media: @CCdiglib; facebook.com/honnoldlibrary; facebook.com/ClaremontCollegesDigitalLibrary; flickr.com/photos/claremontcollegesdigitallibrary

PROGRAM OVERVIEW

Mission/description: The Center for Digital Initiatives facilitates the dissemination of knowledge by providing publishing platforms, consulting, and technical services to enable the creation and distribution of teaching and research resources to the scholarly community.

Year publishing activities began: 2006 (first journal); 2010 (in earnest)

Organization: centralized library publishing unit/department

Staff in support of publishing activities (FTE): library staff (2)

PUBLISHING ACTIVITIES

Types of publications: faculty-driven journals (4); student-driven journals (4); technical/research reports (4); faculty conference papers and proceedings (63); student conference papers and proceedings (106); ETDs (162); undergraduate capstone/honors theses (1182); digital collections; lectures and symposia

Media formats: text; images; video; multimedia/interactive content

Disciplinary specialties: arts and humanities; social and behavioral sciences; physical and mathematical sciences; life sciences; business

Top publications: CMC senior theses; *Journal of Humanistic Mathematics* (journal); *STEAM* (journal); Scripps senior theses; *LUX* (journal); *Performance Practice Review* (journal)

Percentage of journals that are peer reviewed: 100

Campus partners: campus departments or programs; individual faculty; graduate students; undergraduate students

Other partners: Rancho Santa Ana Botanical Gardens

Publishing platform(s): bepress (Digital Commons); CONTENTdm

Digital preservation strategy: Amazon Glacier; Amazon S3; looking into CLOCKSS, LOCKSS, and some others

Additional services: graphic design (print or web); marketing; outreach; training; analytics; cataloging; metadata; ISSN registration; DOI assignment/allocation of identifiers; open URL support; peer review management; business model development; contract/license preparation; author copyright advisory; other author advisory; digitization; hosting of supplemental content; audio/video streaming

Plans for expansion/future directions: Possible expansion into areas of education and alternative/non-traditional publishing.

COLBY COLLEGE

Colby College Libraries

Primary Unit: Digital and Special Collections

Primary Contact:
Marty Kelly
Assistant Director for Digital Collections
207-859-5162
mfkelly@colby.edu

PROGRAM OVERVIEW

Mission/description: The publishing mission of Colby College Libraries Digital and Special Collections is to showcase the scholarly work of Colby's faculty and students, make the college's unique collections more broadly available, and contribute to open intellectual discourse.

Year publishing activities began: 2006

Organization: centralized library publishing unit/department

Staff in support of publishing activities (FTE): library staff (2); undergraduate students (1.5)

Funding sources (%): library operating budget (100)

PUBLISHING ACTIVITIES

Types of publications: faculty-driven journals (1); student-driven journals (1); monographs (14); technical/research reports (1); newsletters (1); undergraduate capstone/honors theses (35); alumni magazine

Media formats: text; images; audio; video

Disciplinary specialties: humanities; environmental science; Jewish studies; economics

Top publications: *Colby Quarterly* (journal); Colby honors theses and Senior Scholars papers; Colby Undergraduate Research Symposium (conference proceedings); *Atlas of Maine* (journal); *Colby Magazine* (magazine)

Percentage of journals that are peer reviewed: 50

Campus partners: campus departments or programs; individual faculty; undergraduate students

Publishing platform(s): bepress (Digital Commons); WordPress

Digital preservation strategy: digital preservation services under discussion

Additional services: outreach; training; analytics; cataloging; metadata; open URL support; peer review management; digitization; hosting of supplemental content; audio/video streaming

Plans for expansion/future directions: Planning to support two new major publishing initiatives with Colby's Center for the Arts and Humanities this coming academic year: the relaunch of the *Colby Quarterly* (1943–2003) and the development of a new undergraduate research journal.

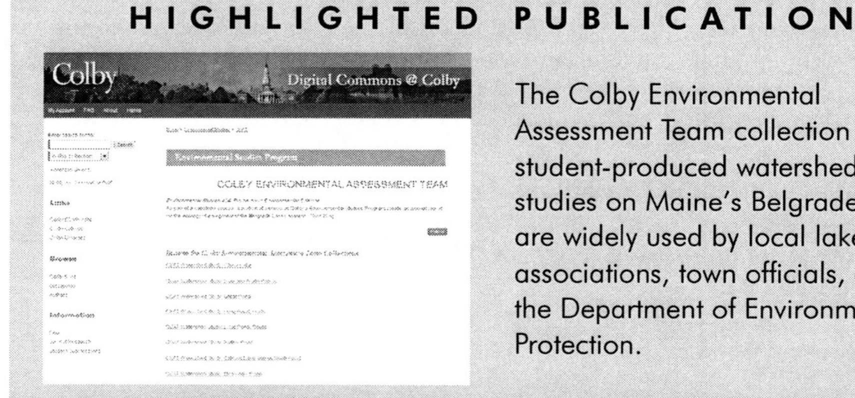

COLLEGE AT BROCKPORT, SUNY

Drake Memorial Library

Primary Unit: Library Technology

Primary Contact:
Kim Myers
Digital Repository Specialist
585-395-2742
kmyers@brockport.edu

PROGRAM OVERVIEW

Year publishing activities began: 2012

Organization: services are distributed across library units/departments

Staff in support of publishing activities (FTE): library staff (1)

Funding sources (%): library operating budget (100)

PUBLISHING ACTIVITIES

Types of publications: faculty-driven journals (1); student-driven journals (2); monographs (1); technical/research reports (140); faculty conference papers and proceedings (7); student conference papers and proceedings (35); newsletters (7); databases (1); ETDs (495); undergraduate capstone/honors theses (85)

Media formats: text; images; audio; video; data

Disciplinary specialties: kinesiology; sports science; physical education; education; counselor education; philosophy; English

Top publications: Counselor Education master's theses; Education master's theses; technical reports from the Water Research Community; *Dissenting Voices* (journal); *Journal of Literary Onomastic Studies* (journal)

Percentage of journals that are peer reviewed: 33

Campus partners: campus departments or programs; individual faculty; graduate students

Publishing platform(s): bepress (Digital Commons)

Digital preservation strategy: LOCKSS

Additional services: copy-editing; marketing; training; cataloging; metadata; ISSN registration; author copyright advisory; digitization; hosting of supplemental content

Plans for expansion/future directions: Expanding in the ETD arena; working with the Graduate School to automate the publication of our master's theses as they are produced.

COLLEGE OF WOOSTER

College of Wooster Libraries

Primary Unit: Digital Scholarship and Services

Primary Contact:
Stephen Flynn
Emerging Technologies Librarian
330-263-2154
sflynn@wooster.edu

PROGRAM OVERVIEW

Mission/description: Our goal is to digitally preserve and promote the original scholarship of our faculty and students.

Year publishing activities began: 2012

Organization: centralized library publishing unit/department

Staff in support of publishing activities (FTE): library staff (1.5); undergraduate students (4)

Funding sources (%): library operating budget (100)

PUBLISHING ACTIVITIES

Types of publications: undergraduate capstone/honors theses (414)

Media formats: text; images; audio; video; data

Campus partners: campus departments or programs; individual faculty; undergraduate students

Publishing platform(s): bepress (Digital Commons); DSpace; WordPress

Digital preservation strategy: in-house; digital preservation services under discussion

Additional services: metadata; digitization; hosting of supplemental content

Plans for expansion/future directions: Migrating from DSpace to bepress, which may enable us to promote the publishing of new undergraduate journals.

COLUMBIA UNIVERSITY

Columbia University Libraries/Information Services

Library Publishing Coalition — Founding Institution

Primary Unit: Center for Digital Research and Scholarship
info@cdrs.columbia.edu

Primary Contact:
Mark Newton
Production Manager
212-851-7337
mnewton@columbia.edu

Website: cdrs.columbia.edu

Social media: @ColumbiaCDRS; @ResearchAtCU; @DataAtCU; @ScholarlyComm; facebook.com/pages/Center-for-Digital-Research-and-Scholarship-Columbia-University/63932011889

PROGRAM OVERVIEW

Mission/description: The Center for Digital Research and Scholarship (CDRS) serves the digital research and scholarly communications needs of the faculty, students, and staff of Columbia University and its affiliates. Our mission is to increase the utility and impact of research produced at Columbia by creating, adapting, implementing, supporting, and sustaining innovative digital tools and publishing platforms for content delivery, discovery, analysis, data curation, and preservation. In pursuit of that mission, we also engage in extensive outreach, education, and advocacy to ensure that the scholarly work produced at Columbia University has a global reach and accelerates the pace of research across disciplines.

Year publishing activities began: 1997 (Columbia University Libraries); 2007 (CDRS)

Organization: centralized library publishing unit/department

Staff in support of publishing activities (FTE): library staff (14.5); graduate students (0.5); undergraduate students (1.25)

Funding sources (%): library operating budget (77); non-library campus budget (3); grants (2); licensing (17); charge backs (1)

PUBLISHING ACTIVITIES

Types of publications: faculty-driven journals (4); student-driven journals (15); journals produced under contract/MOU for external groups (19); monographs (36); technical/research reports (2667); conference papers and proceedings (1150); databases (1); ETDs (1627); undergraduate capstone/honors theses (130)

Media formats: text; images; audio; video; data; software

Disciplinary specialties: law; humanities; public health; global studies; interdisciplinary studies

Top publications: *Tremor and Other Hyperkinetic Movements* (journal); Dangerous Citizens (website); Academic Commons (digital research repository); Women Film Pioneers Project (website); *Columbia Business Law Review* (journal)

Percentage of journals that are peer reviewed: 100

Campus partners: Columbia University Press; campus departments or programs; individual faculty; graduate students; undergraduate students

Other partners: Modern Languages Association; Fordham University Press; New York University; Ecological Society of America. Informal partners include California Digital Library; Cornell University; Purdue University.

Publishing platform(s): Fedora; OJS/OCS/OMP; WordPress; locally developed software; Drupal

Digital preservation strategy: APTrust; Archive-It; DuraCloud/DSpace; DPN; in-house; digital preservation services under discussion. Content is also backed up to NYSERnet, to two on-site locations, and off-site to tape with IronMountain.

Additional services: graphic design (print or web); typesetting; copyediting; marketing; outreach; training; analytics; cataloging; metadata; ISSN registration; DOI assignment/allocation of identifiers; open URL support; dataset

HIGHLIGHTED PUBLICATION

COLUMBIA UNIVERSITY
ACADEMIC COMMONS

Academic Commons is a digital publication platform that brings global visibility to the research and scholarship of Columbia University and its affiliates.

academiccommons.columbia.edu

management; business model development; contract/license preparation; author copyright advisory; other author advisory; digitization; hosting of supplemental content; audio/video streaming; preservation; repository deposit to PMC; SEO; application development; content and platform migration; workshops and consultation; social media and journal publishing best practices workshops; informal scholarly communication events; Open Access Week events; Campus OA fund management; collaboration spaces

Plans for expansion/future directions: Planning to continue integration of the publishing program with the digital research repository, Academic Commons (academiccommons.columbia.edu), as well as to pursue new publishing partnerships with scholarly societies through members affiliated with the university. Further plans include expansion into unique identifier support (such as with ORCID and through EZID) as well as work in support of federal and funder mandates for access to funded research.

CONNECTICUT COLLEGE

Charles E. Shain Library

Primary Unit: Special Collections

Primary Contact:
Benjamin Panciera
Director of Special Collections
860-439-2654
bpancier@conncoll.edu

Website: digitalcommons.conncoll.edu

PROGRAM OVERVIEW

Mission/description: Connecticut College seeks to make the products of student and faculty research and campus resources as widely available as possible through its institutional repository. Mandatory electronic submission of student honors theses began in 2011. The faculty overwhelmingly passed an open access policy in 2013, and the library has supported this by retrospectively making faculty research available through the institutional repository.

Year publishing activities began: 2006

Organization: centralized library publishing unit/department

Staff in support of publishing activities (FTE): library staff (1.5)

Funding sources (%): library operating budget (100)

PUBLISHING ACTIVITIES

Types of publications: faculty conference papers and proceedings (20); newsletters (3); undergraduate capstone/honors theses (60)

Media formats: text; audio

Campus partners: campus departments or programs; individual faculty; undergraduate students

Publishing platform(s): bepress (Digital Commons)

Digital preservation strategy: digital preservation services under discussion; no digital preservation services provided

Additional services: cataloging; metadata; author copyright advisory; digitization; hosting of supplemental content; audio/video streaming

Plans for expansion/future directions: Seeking to optimize faculty participation and maximize the amount of available research in the institutional repository and inform faculty of the possibility of using the repository to make available unpublished material like conference papers and datasets.

CORNELL UNIVERSITY

Cornell University Library

Primary Unit: Digital Scholarship and Preservation Services

Primary Contact:
David Ruddy
Director, Scholarly Communications Services
607-255-6803
dwr4@cornell.edu

PROGRAM OVERVIEW

Mission/description: Separate operations have their own mission statements (Project Euclid, arXiv, eCommons, CIP). In general, we wish to promote sustainable models of scholarly communications with an emphasis on access and affordability.

Year publishing activities began: 2000

Organization: services are primarily distributed across library units. A few projects involve the Cornell University Press.

Staff in support of publishing activities (FTE): library staff (7); undergraduate students (0.3)

Funding sources (%): library operating budget (20); sales revenue (40); other (40)

PUBLISHING ACTIVITIES

Types of publications: faculty-driven journals (3); student-driven journals (1); journals produced under contract/MOU for external groups (69); monographs (170); technical/research reports (85000); ETDs (700); undergraduate capstone/ honors theses (50); case studies

Media formats: text; audio; video; data

Disciplinary specialties: mathematics; physics; statistics; computer science

Percentage of journals that are peer reviewed: 100

Campus partners: Cornell University Press; campus departments or programs; individual faculty; graduate students

Other partners: Duke University Press; scholarly societies; scholars worldwide

Publishing platform(s): DPubS; DSpace; locally developed software

Digital preservation strategy: in-house

Additional services: graphic design (print or web); metadata; DOI assignment/allocation of identifiers; open URL support; budget preparation; digitization; hosting of supplemental content; audio/video streaming

Additional information: "Publishing" activities at Cornell are complex and include at least four fairly distinct operations: Project Euclid, arXiv.org, eCommons (an institutional repository), and Cornell Initiatives in Publishing (Cornell-related journals and books). Each of these operations arguably fit the provided criteria for "library publishing" activities.

DARTMOUTH COLLEGE

Dartmouth College Library

Library Publishing Coalition

Primary Unit: Digital Library Program
library.dartmouth.edu/mail/send.php?to=askalib

Primary Contact:
Elizabeth Kirk
Associate Librarian for Information Resources
603-646-9929
elizabeth.e.kirk@dartmouth.edu

Website: www.dartmouth.edu/~library/digital

PROGRAM OVERVIEW

Mission/description: The Dartmouth College Library's Digital Publishing Program supports faculty publication of original scholarly content in a digital environment. Our digital publications include journals, monographs, and scholarly editions. All content is available online without charge.

Year publishing activities began: 2002

Organization: services are distributed across library units/departments

Staff in support of publishing activities (FTE): library staff (3.75)

Funding sources (%): library operating budget (10); endowment income (10); other (80)

PUBLISHING ACTIVITIES

Types of publications: faculty-driven journals (3); monographs (1); ETDs (400); digital, scholarly editions of manuscripts, letters, etc. (4)

Media formats: text; images; audio; video; data; concept maps/modeling maps/ visualizations; multimedia/interactive content

Disciplinary specialties: environment; linguistics; electronic or "new" media; Native American history; history of Arctic exploration

Top publications: *Elementa* (journal); *Linguistic Discovery* (journal); *Journal of E-Media Studies* (journal); Occom Circle Project (digital collection); *Artistry of the Homeric Simile* (monograph)

Percentage of journals that are peer reviewed: 100

Campus partners: campus departments or programs; individual faculty

Other partners: University Press of New England; BioOne

Publishing platform(s): CONTENTdm; locally developed software; Ambra

Digital preservation strategy: DPN; HathiTrust; LOCKSS; Portico; in-house; digital preservation services under discussion

Additional services: marketing; outreach; training; analytics; cataloging; metadata; ISSN registration; DOI assignment/allocation of identifiers; open URL support; peer review management; business model development; budget preparation; other author advisory; digitization; audio/video streaming; XML consultation in JATS 1.0 and TEI

Additional information: The partnership with the publisher BioOne is enabling us to increase our technological capacity for journal publishing. BioOne is a significant contributor to the staffing for *Elementa*. The partnership with the University Press of New England is enabling us to increase knowledge and capacity for monograph publishing.

Plans for expansion/future directions: Publishing more monographs in conjunction with the University Press of New England, further developing technical capacity for journals, increasing the number of digital editions, working with student journals.

HIGHLIGHTED PUBLICATION

Through *Elementa: Science of the Anthropocene,* we aim to facilitate scientific solutions to the challenges presented by this era of accelerated human impact with timely, technically sound, peer-reviewed articles that address interactions between human and natural systems and behaviors.

home.elementascience.org

DUKE UNIVERSITY

Duke University Libraries

Library Publishing Coalition

Primary Unit: Office of Copyright and Scholarly Communications
open-access@duke.edu

Primary Contact:
Paolo Mangiafico
Coordinator of Scholarly Communications Technology
919-613-6317
paolo.mangiafico@duke.edu

Website: library.duke.edu/openaccess

PROGRAM OVERVIEW

Mission/description: Duke University Libraries partners with members of the Duke community to publish and disseminate scholarship in new and creative ways, including helping to publish scholarly journals on an open access digital platform, archiving previously published and original works, and consulting on new forms of scholarly dissemination.

Year publishing activities began: 2007

Organization: services are distributed across library units/departments

Staff in support of publishing activities (FTE): library staff (1.5); graduate students (0.5)

Funding sources (%): library operating budget (100)

PUBLISHING ACTIVITIES

Types of publications: faculty-driven journals (3); technical/research reports (11); databases (2); ETDs (500); undergraduate capstone/honors theses (40)

Media formats: text; images; audio; data; multimedia/interactive content

Disciplinary specialties: Greek, Roman, and Byzantine studies; transatlantic German studies; 18th-century Russian studies; cultural anthropology; scholarly communications

Top publications: *Cultural Anthropology* (journal); ETDs; *Greek, Roman, and Byzantine Studies* (journal); Scholarly Communications @ Duke (blog); *Andererseits* (journal)

Percentage of journals that are peer reviewed: 100

Campus partners: campus departments or programs; individual faculty; graduate students; undergraduate students

Other partners: Society for Cultural Anthropology; editors of particular journals and their organizations

Publishing platform(s): DSpace; OJS; WordPress; Symplectic Elements

Digital preservation strategy: depends on the journal and type of content, primarily in-house, but exploring archiving with Portico

Additional services: outreach; training; analytics; metadata; open URL support; dataset management; business model development; contract/license preparation; author copyright advisory; other author advisory; hosting of supplemental content

Plans for expansion/future directions: Working with more datasets, digital projects, and forms other than linear text; exploring platforms that support new publishing models, not just digital versions of old journal models.

HIGHLIGHTED PUBLICATION

Cultural Anthropology is the journal of the Society for Cultural Anthropology, a section of the American Anthropological Association (AAA). It is one of 22 journals published by the AAA, and it is widely regarded as one of the flagship journals of its discipline.

culanth.org

EMORY UNIVERSITY

Robert W. Woodruff Library

Primary Unit: Emory Center for Digital Scholarship
allen.tullos@emory.edu

Primary Contact:
Stewart Varner
Digital Scholarship Coordinator
404-727-1196
stewart.varner@emory.edu

PROGRAM OVERVIEW

Mission/description: The enduring goal of a university is to create and disseminate knowledge. Changes in technology offer opportunities for new forms of both creation and dissemination of scholarship through open access (OA). Open access publishing also offers opportunities for Emory University to fulfill its mission of creating and preserving knowledge in a way that opens disciplinary boundaries and facilitates sharing that knowledge more freely with the world.

Year publishing activities began: 1995

Organization: services are distributed across library units/departments

Staff in support of publishing activities (FTE): library staff (4); graduate students (6)

Funding sources (%): library operating budget (95); grants (5)

PUBLISHING ACTIVITIES

Types of publications: faculty-driven journals (4); student-driven journals (1); databases (18); ETDs (758)

Media formats: text; images; audio; video; concept maps/modeling maps/ visualizations; multimedia/interactive content

Disciplinary specialties: Southern studies; religion/theology

Top publications: *Southern Spaces* (journal); *Molecular Vision* (journal); *Methodist Review* (journal); *Practical Matters* (journal)

Percentage of journals that are peer reviewed: 100

Campus partners: campus departments or programs; individual faculty; graduate students

Publishing platform(s): Fedora; OJS/OCS/OMP; WordPress; Drupal

Digital preservation strategy: digital preservation services under discussion

Additional services: typesetting; copy-editing; metadata; peer review management; contract/license preparation; author copyright advisory; digitization; audio/video streaming

Additional Information: On June 1, 2013, Emory University announced the launch of the Emory Center for Digital Scholarship (ECDS). The ECDS brings together four units currently housed in the Robert W. Woodruff Library: the Digital Scholarship Commons (DiSC), the Electronic Data Center, the Lewis H. Beck Center for Electronic Collections, and the Emory Center for Interactive Teaching (ECIT). These units have each collaborated with Emory scholars who wish to incorporate technology into their teaching and research. The formation of the ECDS will break down barriers between these functions and simplify the process of establishing partnerships with scholars. Expanding and strengthening support for open access, digital publishing is a top priority for the ECDS.

Plans for expansion/future directions: Reexamining the expansion of library publishing services following the recent launch of the Emory Center for Digital Scholarship.

FLORIDA ATLANTIC UNIVERSITY

SE Wimberly Library

Primary Unit: Digital Library
lydig@fau.edu

Primary Contact:
Joanne Parandjuk
Digital Initiatives Librarian
561-297-0139
jparandj@fau.edu

Website: www.library.fau.edu/depts/digital_library/about.htm

PROGRAM OVERVIEW

Mission/description: Recognizing the publishing needs of campus members and local partners, an open access publishing service was initiated by the FAU Digital Library in support of scholarly communications across campus and the wider dissemination of FAU research and creative content.

Year publishing activities began: 2011

Organization: centralized library publishing unit/department

Staff in support of publishing activities (FTE): library staff (1); graduate students (0.5)

Funding sources (%): library operating budget (85); charge backs (15)

PUBLISHING ACTIVITIES

Types of publications: faculty-driven journals (3); student-driven journals (1); journals produced under contract/MOU for external groups (1); ETDs (185); undergraduate capstone/honors theses (50)

Media formats: text; images; audio; video

Disciplinary specialties: geosciences; undergraduate research; communications; local history

Top publications: *The Florida Geographer* (journal); *Democratic Communique* (journal); *FAU Undergraduate Research Journal* (journal); *Journal of Coastal Research* (journal backfile); *Broward Legacy* (journal)

Percentage of journals that are peer reviewed: 80

Campus partners: campus departments or programs; individual faculty; undergraduate students

Other partners: Broward County Historical Society

Publishing platform(s): Islandora (migration underway); OJS/OCS/OMP

Digital preservation strategy: Florida Digital Archive member

Additional services: graphic design (print or web); marketing; outreach; training; analytics; cataloging; metadata; ISSN registration; digitization; audio/video streaming

Plans for expansion/future directions: We have just launched the second issue of the first volume of our *Undergraduate Research Journal* to instill scholarly inquiry and practices among undergraduates, and we hope to see a rise in the research activity of our students.

FLORIDA STATE UNIVERSITY

Robert Manning Strozier Library

Library Publishing Coalition

Primary Unit: Technology and Digital Scholarship

Primary Contact:
Micah Vandegrift
Scholarly Communication Librarian
850-645-9756
mvandegrift@fsu.edu

Website: diginole.lib.fsu.edu

PROGRAM OVERVIEW

Mission/description: Scholarly communications is a developing area of librarianship that deals with the production, dissemination, promotion, and preservation of scholarly research and creative works. The Scholarly Communication initiative will find, assess, and provide tools and services for representing scholarship in a digital environment. Our vision is to support a variety of new modes and models of dissemination for academic work (open access, digital publishing, project-based digital scholarship, etc.). Areas of focus include our institutional repository (technical management, outreach, collection development); Open Access (education and programs on access options for scholarly work); author rights (information and resources on negotiating copyright transfer contracts); copyrights and fair use (information and resources on copyright as it pertains to academic publishing); research and writing (keeping abreast of the many changes and development in this area, and contributing to the professional literature); and outreach (creating partnerships with campus offices, faculty, and administrators to further the scholarly communications initiative).

Year publishing activities began: 2011

Organization: centralized library publishing unit/department

Staff in support of publishing activities (FTE): library staff (1); graduate students (1)

Funding sources (%): library operating budget (100)

PUBLISHING ACTIVITIES

Types of publications: faculty-driven journals (3); student-driven journals (1); monographs (1); technical/research reports (1); newsletters (1); ETDs (1000); undergraduate capstone/honors theses (100)

Media formats: text

Disciplinary specialties: arts and literature; art education and therapy; law

Top publications: *HEAL: Humanism Evolving through Arts and Literature* (journal); *Journal of Art for Life* (journal); *The Owl* (journal); *FSU Law Review* (journal)

Campus partners: campus departments or programs; individual faculty; graduate students; undergraduate students

Publishing platform(s): bepress (Digital Commons)

Digital preservation strategy: digital preservation services under discussion

Additional services: marketing; outreach; training; analytics; metadata; ISSN registration; peer review management; contract/license preparation; author copyright advisory; hosting of supplemental content

Plans for expansion/future directions: Piloting an Open Access fund, finding a sustainable model and including it as an ongoing resource for moving scholarship and prestige to Open Access; growing ScholComm office to include Repository Manager and host research fellows (CLIR, Mellon); coordinating with the School of Library and Information Studies and the History of Text Technologies to integrate ScholComm initiatives into curriculum; providing training and investment in FSU LIS students' skills and knowledge in this area; reworking open access policy with Faculty Senate to make our policy more effective and more in line with the scholarly communication push internationally.

GEORGETOWN UNIVERSITY

Georgetown University Libraries

Primary Unit: Library Information Technologies
digitalscholarship@georgetown.edu

Primary Contact:
Kate Dohe
Digital Services Librarian
202-687-6387
kd602@georgetown.edu

Website: www.library.georgetown.edu/digitalgeorgetown

Social media: @gtownlibrary

PROGRAM OVERVIEW

Mission/description: DigitalGeorgetown supports the advancement of education and scholarship at Georgetown and contributes to the expansion of research initiatives, both nationally and internationally. By providing the infrastructure, resources, and services, DigitalGeorgetown sustains the evolution from the traditional research models of today to the enriched scholarly communication environment of tomorrow, and it provides context and leadership in developing collaborative opportunities with partners across the campus and around the world.

Year publishing activities began: 2009

Organization: centralized library publishing unit/department

Staff in support of publishing activities (FTE): library staff (1.5)

Funding sources (%): library operating budget (100)

PUBLISHING ACTIVITIES

Types of publications: student-driven journals (2); monographs (69); technical/ research reports (23); newsletters (119); ETDs (2428); undergraduate capstone/ honors theses (246); faculty papers; video interviews; citations; syllabi

Media formats: text; images; audio; video; data; concept maps/modeling maps/ visualizations; multimedia/interactive content

Disciplinary specialties: linguistics; communications; international relations/ foreign policy; bioethics

Top publications: *Georgetown University Round Tables on Language and Linguistics 1999* (monograph); *The Human Cloning Debate* (monograph); *The Genocide in Cambodia* (monograph)

Percentage of journals that are peer reviewed: 0

Campus partners: Georgetown University Press; campus departments or programs; individual faculty; graduate students; undergraduate students

Publishing platform(s): DSpace

Digital preservation strategy: digital preservation services under discussion

Additional services: marketing; outreach; training; analytics; cataloging; metadata; contract/license preparation; author copyright advisory; other author advisory; digitization; hosting of supplemental content; audio/video streaming

Plans for expansion/future directions: Continuing to enhance and expand our initiative to include more open access materials, different forms and formats of ETDs, and other scholarly publications.

GEORGIA STATE UNIVERSITY

Georgia State University Library

Primary Unit: Digital Initiatives
digitalarchive@gsu.edu

Primary Contact:
Sean Lind
Digital Initiatives Librarian
404-413-2757
slind1@gsu.edu

Website: digitalarchive.gsu.edu

PROGRAM OVERVIEW

Mission/description: The mission of the institutional repository at Georgia State University is to give free and open access to the impactful scholarly and creative works, research, publications, reports, and data contributed by faculty, students, staff, and administrative units of Georgia State University.

Year publishing activities began: 2009

Organization: services are distributed across campus

Staff in support of publishing activities (FTE): library staff (1)

Funding sources (%): library materials budget (100)

PUBLISHING ACTIVITIES

Types of publications: student-driven journals (3); monographs (1); faculty conference papers and proceedings (10); newsletters (3); ETDs (510); undergraduate capstone/honors theses (13)

Media formats: text

Disciplinary specialties: law review; undergraduate honors research

Top publications: *Georgia State University Law Review* (journal); *Colonial Academic Alliance Undergraduate Research Journal* (journal); *DISCOVERY: Georgia State University Undergraduate Honors Research Journal* (journal)

Percentage of journals that are peer reviewed: 100

Campus partners: campus departments or programs; individual faculty; undergraduate students

Publishing platform(s): bepress (Digital Commons)

Digital preservation strategy: DuraCloud/DSpace

Additional services: marketing; outreach; training; analytics; cataloging; metadata; ISSN registration; open URL support; author copyright advisory; other author advisory; digitization

Plans for expansion/future directions: Increasing the number and variety of Georgia State University faculty scholarly publications openly available for download on the Internet.

GRAND VALLEY STATE UNIVERSITY

Grand Valley State University Libraries

Primary Unit: Collections and Scholarly Communications
scholarworks@gvsu.edu

Primary Contact:
Sarah Beaubien
Scholarly Communications Outreach Coordinator
616-331-2631
beaubisa@gvsu.edu

PROGRAM OVERVIEW

Year publishing activities began: 2008

Organization: centralized library publishing unit/department

Staff in support of publishing activities (FTE): library staff (1.75)

Funding sources (%): library operating budget (100)

PUBLISHING ACTIVITIES

Types of publications: faculty-driven journals (6); student-driven journals (5); textbooks (5); technical/research reports (8); faculty conference papers and proceedings (29); newsletters (2); ETDs (64); undergraduate capstone/honors theses (54)

Media formats: text; images; audio; video; data

Top publications: *Online Readings in Psychology and Culture* (digital collection); *Foundation Review* (journal); *Fishladder* (journal); *Language Arts Journal of Michigan* (journal); *Journal of Tourism Insights* (journal)

Percentage of journals that are peer reviewed: 35

Campus partners: campus departments or programs; individual faculty; graduate students; undergraduate students

Other partners: Michigan Council of Teachers of English; Resort and Commercial Recreation Association; International Association for Cross-Cultural Psychology; Johnson Center for Philanthropy

Publishing platform(s): bepress (Digital Commons)

Digital preservation strategy: LOCKSS; Portico; digital preservation services under discussion

Additional services: outreach; training; analytics; cataloging; metadata; ISSN registration; DOI assignment/allocation of identifiers; peer review management; author copyright advisory; digitization; hosting of supplemental content

HIGHLIGHTED PUBLICATION

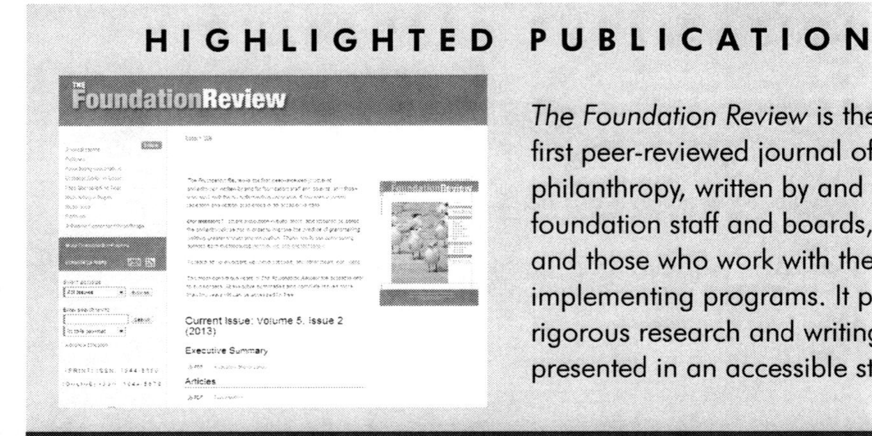

The Foundation Review is the first peer-reviewed journal of philanthropy, written by and for foundation staff and boards, and those who work with them implementing programs. It provides rigorous research and writing, presented in an accessible style.

scholarworks.gvsu.edu/tfr

GUSTAVUS ADOLPHUS COLLEGE

Folke Bernadotte Memorial Library

Primary Contact:
Barbara Fister
Professor and Academic Librarian
507-933-7553
fister@gac.edu

PROGRAM OVERVIEW

Mission/description: We want to support the shift from closed, licensed access to information to open, shareable, and sustainable scholarship.

Year publishing activities began: 2012

Organization: entrepreneurial, experimental, more or less a sandbox in which librarians help other faculty consider alternatives

Staff in support of publishing activities (FTE): library staff (0.1)

Funding sources (%): other (100)

PUBLISHING ACTIVITIES

Types of publications: monographs (1)

Media formats: text

Campus partners: individual faculty

Publishing platform(s): CONTENTdm; WordPress; PressBooks

Digital preservation strategy: in-house

Additional services: author copyright advisory; other author advisory; digitization

Additional information: We have published one monograph using PressBooks: an anthology based on faculty statements about teaching, scholarship, and service submitted for tenure and promotion. We wanted it to be lightweight and without cost other than time. It worked. We also have shared platform advice with faculty interested in publishing. It is all very much at the beginning and is without much in the way of technical or financial support, but we expect the resource commitment to grow.

Plans for expansion/future directions: Working with similar libraries to study the possible launch of a press.

HAMILTON COLLEGE

Burke Library

Primary Unit: Department of Special Collections and Archives
cgoodwil@hamilton.edu

Primary Contact:
Randall Ericson
Editor
503-894-9804
rericson@hamilton.edu

Website: couperpress.org

PROGRAM OVERVIEW

Mission/description: The Couper Press was established in 2006 by Couper Librarian Randall Ericson of the Burke Library at Hamilton College in Clinton, New York. The press is named in honor of the late Richard W. Couper '44, an alumnus, life trustee of Hamilton, and benefactor of the Burke Library. The press publishes a quarterly journal of scholarship, *American Communal Societies Quarterly* (*ACSQ*), which showcases the communal societies collections of Burke Library. American Communal Societies Series, a monograph series, presents new scholarship pertaining to American intentional communities as well as reprints of, and critical introductions to, important historical works that may be difficult to find or are out of print. Shaker Studies are short monographs on the Shakers. Occasional publications are published on topics that highlight the special collections of the Burke Library.

Year publishing activities began: 2006

Organization: centralized library publishing unit/department

Staff in support of publishing activities (FTE): library staff (1.25)

Funding sources (%): endowment income (100)

PUBLISHING ACTIVITIES

Types of publications: faculty-driven journals (4); monographs (5)

Media formats: text; images

Disciplinary specialties: communal studies; religious studies; sociology; American history; musicology

Top publications: *Prison Diary and Letters of Chester Gillette* (monograph); *Visiting the Shakers, 1778–1849: Watervliet, Hancock, Tyringham, New Lebanon* (monograph); *Encyclopedic Guide to American Intentional Communities* (monograph); *A Promising Venture: Shaker Photographs from the WPA* (monograph); *Demographic Directory of the Harmony Society* (monograph)

Campus partners: individual faculty; undergraduate students

Other partners: museums; libraries; private collectors

Digital preservation strategy: digital preservation services under discussion

Additional services: graphic design (print or web); copy-editing

Plans for expansion/future directions: Considering making the *American Communal Societies Quarterly* available through an institutional repository.

ILLINOIS WESLEYAN UNIVERSITY

The Ames Library

Primary Unit: Scholarly Communications

Primary Contact:
Stephanie Davis-Kahl
Scholarly Communications Librarian
309-556-3010
sdaviska@iwu.edu

PROGRAM OVERVIEW

Mission/description: The Ames Library publishing program focuses on disseminating excellent student-authored research, scholarship, and creative works, with an emphasis on providing education and outreach on issues related to publishing such as open access, author rights, and copyright.

Year publishing activities began: 2008

Organization: centralized library publishing unit/department

Staff in support of publishing activities (FTE): library staff (2); undergraduate students (3)

Funding sources (%): library operating budget (25); non-library campus budget (75)

PUBLISHING ACTIVITIES

Types of publications: student-driven journals (6); textbooks; (1); student conference papers and proceedings (1); newsletters (1); undergraduate capstone/ honors theses (25)

Media formats: text; images; audio; video

Disciplinary specialties: economics; political science; history

Top publications: *Undergraduate Economic Review* (journal); *Constructing History* (journal); *Res Publica* (journal)

Percentage of journals that are peer reviewed: 100

Campus partners: campus departments or programs; individual faculty; undergraduate students

Publishing platform(s): bepress (Digital Commons)

Digital preservation strategy: in-house; digital preservation services under discussion

Additional services: training; analytics; metadata; peer review management; author copyright advisory; other author advisory; hosting of supplemental content; audio/video streaming

Additional information: Regarding our funding model; 25 percent of the cost of our bepress implementation is covered by the library, while the remaining 75 percent is generously provided by the Office of the President, Office of the Provost, and Mellon Center for Faculty and Curriculum Development. Faculty advisors for our student journals donate their time.

Plans for expansion/future directions: Considering how to best position the program to become a publishing outlet for faculty.

INDIANA UNIVERSITY

Indiana University Libraries

Primary Unit: IUScholarWorks
iusw@indiana.edu

Primary Contact:
Jennifer Laherty
Digital Publishing Librarian
812-855-5609
jlaherty@indiana.edu

Website: scholarworks.iu.edu

PROGRAM OVERVIEW

Mission/description: IUScholarWorks is a set of services from the Indiana University Libraries to make the work of IU scholars freely available and to ensure that these resources are preserved and organized for the future.

Year publishing activities began: 2006

Organization: services are distributed across library units/departments

Staff in support of publishing activities (FTE): library staff (3.5); graduate students (1)

Funding sources (%): library materials budget (20); library operating budget (80)

PUBLISHING ACTIVITIES

Types of publications: faculty-driven journals (14); student-driven journals (2); technical/research reports (2); newsletters (1); ETDs (300)

Media formats: text; images; audio; video; data; concept maps/modeling maps/visualizations; multimedia/interactive content

Disciplinary specialties: folklore

Percentage of journals that are peer reviewed: 90

Campus partners: IU Press; campus departments or programs; individual faculty; graduate students; undergraduate students

Other partners: American Folklore Society

Publishing platform(s): DSpace; OJS/OCS/OMP

Digital preservation strategy: Archive-It; CLOCKSS; DuraCloud/DSpace; HathiTrust

Additional services: outreach; training; analytics; cataloging; notification of A&I sources; ISSN registration; DOI assignment/allocation of identifiers; dataset management; peer review management; author copyright advisory; digitization; hosting of supplemental content; audio/video streaming; metadata consultation.

Plans for expansion/future directions: Incorporating the Libraries' open access publishing activities into the development of a new campus office, the Office of Scholarly Publishing, which includes the University Press and an eTextbook initiative.

JOHNS HOPKINS UNIVERSITY

Sheridan Libraries

Primary Unit: Scholarly Resources and Special Collections
dissertations@jhu.edu

Primary Contact:
David Reynolds
Manager of Scholarly Digital Initiatives
410-516-7220
davidr@jhu.edu

PROGRAM OVERVIEW

Mission/description: To provide a publishing platform for required ETDs and journals for the Johns Hopkins academic community.

Year publishing activities began: 2009

Organization: services are distributed across library units/departments

Staff in support of publishing activities (FTE): library staff (2.25)

Funding sources (%): library operating budget (100)

PUBLISHING ACTIVITIES

Types of publications: ETDs (5); undergraduate capstone/honors theses (50)

Media formats: text; images

Disciplinary specialties: education; business

Top publications: *International Journal of Interdisciplinary Education* (journal); *New Horizons for Education* (journal)

Percentage of journals that are peer reviewed: 50

Campus partners: campus departments or programs; individual faculty; graduate students

Publishing platform(s): DSpace; OJS/OCS/OMP

Digital preservation strategy: in-house; digital preservation services under discussion

Additional services: training; analytics; metadata; peer review management; author copyright advisory

Additional information: We have only done an ETD pilot so far, but mandatory submission was required as of September 1, 2013. We are working with the School of Education to publish two new OA journals. We expect the inaugural issues to appear by the second quarter of 2014.

Plans for expansion/future directions: Publishing journals for the School of Education; looking into providing a monograph publishing service for academic departments; revisiting the question of publishing student journals.

KANSAS STATE UNIVERSITY

Kansas State University Libraries

Primary Unit: Scholarly Communications and Publishing
info@newprairiepress.org

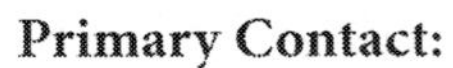

Primary Contact:
Char Simser
Coordinator of Electronic Publishing, New Prairie Press
785-532-7444
info@newprairiepress.org

Website: newprairiepress.org

Social media: @NewPrairiePress

PROGRAM OVERVIEW

Mission/description: To host peer-reviewed scholarly journals, monographs, conference proceedings, and other series primarily in the humanities and social sciences; make the content freely available worldwide; and contribute to and support evolving scholarly publishing models.

Year publishing activities began: 2007

Organization: centralized library publishing unit/department

Staff in support of publishing activities (FTE): library staff (0.7); graduate students (0.3)

Funding sources (%): library operating budget (100)

PUBLISHING ACTIVITIES

Types of publications: faculty-driven journals (7)

Media formats: text; images; video

Disciplinary specialties: financial therapy; rural research and policy; library science; cognitive sciences and semantics; analytical philosophy

Top publications: *GDR Bulletin* (journal); *Baltic International Yearbook* (journal); *Journal of Financial Therapy* (journal); *Online Journal of Rural Research & Policy* (journal); *Kansas Library Association College and University Libraries Section Proceedings* (conference proceedings)

Percentage of journals that are peer reviewed: 100

Campus partners: campus departments or programs; individual faculty

Publishing platform(s): bepress (Digital Commons)

Digital preservation strategy: CLOCKSS

Additional services: graphic design (print or web); marketing; training; notification of A&I sources; DOI assignment/allocation of identifiers; digitization; hosting of supplemental content

Plans for expansion/future directions: Publishing open access monographs and conference proceedings and publishing two undergraduate research journals; setting up an advisory board to help set direction and policy and recommend new titles for NPP.

HIGHLIGHTED PUBLICATION

Since 2010, *The Journal of Financial Therapy* has been the leading forum dedicated to clinical, experimental, and qualitative research in the emerging field of financial therapy.

jftonline.org

LOYOLA UNIVERSITY CHICAGO

Loyola University Chicago Libraries

Primary Unit: Library Systems

Primary Contact:
Margaret Heller
Digital Services Librarian
773-508-2686
mheller1@luc.edu

PROGRAM OVERVIEW

Mission/description: Loyola eCommons is an open-access, sustainable, and secure resource created to preserve and provide access to research, scholarship, and creative works created by the university community for the benefit of Loyola students, faculty, staff, and the larger academic community. Sponsored by the University Libraries, Loyola eCommons is a suite of online resources, services, and people working in concert to facilitate a wide range of scholarly and archival activities, including collaboration, resource sharing, author rights management, digitization, preservation, and access by a global academic audience.

Year publishing activities began: 2011

Organization: centralized library publishing unit/department

Staff in support of publishing activities (FTE): library staff (1); graduate students (0.25); undergraduate students (1.5)

Funding sources (%): library operating budget (90); non-library campus budget (10)

PUBLISHING ACTIVITIES

Types of publications: faculty conference papers and proceedings (182); student conference papers and proceedings (1); ETDs (1996)

Media formats: text; images; data

Disciplinary specialties: criminal justice; economics; social work

Campus partners: campus departments or programs; individual faculty; graduate students; undergraduate students

Publishing platform(s): bepress (Digital Commons)

Digital preservation strategy: digital preservation services under discussion

Additional services: outreach; training; analytics; metadata; digitization; hosting of supplemental content

Plans for expansion/future directions: Hosting conference proceedings and journals.

MACALESTER COLLEGE

DeWitt Wallace Library

Primary Unit: Digital Scholarship and Services

Primary Contact:
Johan Oberg
Digital Scholarship and Services Librarian
651-696-6003
joberg@macalester.edu

Website: www.macalester.edu/library/digitalinitiatives/index.html

PROGRAM OVERVIEW

Mission/description: The Digital Publishing Unit of the DeWitt Wallace Library supports the creation, management, and dissemination of local digital-born scholarship in various formats. Essential to supporting this mission is the continuing exploration of evolving creation, collaboration, and publication tools; encoding methods; and development of staff skills and facility resources. The Unit serves the digital scholarship and electronic publishing needs through development of digital scholarship projects as well as open access online distribution of journals, articles, and conference proceedings. The Library is committed to playing an active role in the changing landscape of scholarly publishing and supports the ideals of the open access movement.

Year publishing activities began: 2004

Organization: services are distributed across library units/departments

Staff in support of publishing activities (FTE): library staff (3)

Funding sources (%): library operating budget (100)

PUBLISHING ACTIVITIES

Types of publications: faculty-driven journals (6); student-driven journals (2); journals produced under contract/MOU for external groups (1); monographs (1); undergraduate capstone/honors theses (100); college alumni magazine; conference proceedings; oral histories

Media formats: text; images; audio; video; data

Disciplinary specialties: natural sciences; social sciences; fine arts; humanities; interdisciplinary studies

Top publications: "An Analysis of the Career Length of Professional Basketball Players" (thesis); "The Cultural Omnivore in Its Natural Habitat: Music Taste at a Liberal Arts College" (thesis); "What are the Effects of Mergers in the U.S. Airline Industry? An Econometric Analysis on Delta-Northwest Merger" (thesis); "The Mirror's Reflection: Virgil's Aeneid in English Translation" (thesis); "Fat Teen Trouble: A Sociological Perspective of Obesity in Adolescents" (thesis)

Percentage of journals that are peer reviewed: 100

Campus partners: campus departments or programs; individual faculty; undergraduate students

Other partners: Association for Nepal and Himalayan Studies (ANHS)

Publishing platform(s): bepress (Digital Commons); CONTENTdm

Digital preservation strategy: in-house

Additional services: typesetting; cataloging; metadata; ISSN registration; dataset management; author copyright advisory; other author advisory; digitization; hosting of supplemental content; audio/video streaming

Plans for expansion/future directions: Working with faculty to develop data management curation and preservation.

MCGILL UNIVERSITY

McGill University Library

Primary Unit: eScholarship, ePublishing and Digitization

Primary Contact:
Amy Buckland
eScholarship, ePublishing & Digitization Coordinator
514-398-3059
amy.buckland@mcgill.ca

PROGRAM OVERVIEW

Mission/description: McGill University Library showcases the research done by the McGill community to the world via publishing initiatives such as electronic theses and dissertations, open access journals and monographs, and by partnering with others to develop new methods to disseminate research.

Year publishing activities began: 1988

Organization: centralized library publishing unit/department

Staff in support of publishing activities (FTE): library staff (0.5)

Funding sources (%): library operating budget (100)

PUBLISHING ACTIVITIES

Types of publications: faculty-driven journals (2); monographs (1); technical/research reports (20); ETDs (1000); undergraduate capstone/honors theses (25); working papers

Media formats: text; images; audio; video

Disciplinary specialties: education; food cultures; library history

Top publications: *McGill Journal of Education* (journal); *CuiZine* (journal); *Fontanus* (journal)

Percentage of journals that are peer reviewed: 100

Campus partners: campus departments or programs; individual faculty; graduate students; undergraduate students

Other partners: Public Knowledge Project; Erudit; ThesesCanada

Publishing platform(s): OJS/OCS/OMP; locally developed software; DigiTool

Digital preservation strategy: in-house; digital preservation services under discussion

Additional services: training; analytics; notification of A&I sources; ISSN registration; author copyright advisory; other author advisory; digitization; hosting of supplemental content

Plans for expansion/future directions: ETD program and *Fontanus* series are well established, but OJS journals are still in a developmental stage; looking to pair with the digital humanities community on campus to look at new ways of publishing, beyond the journal/monograph binary.

MIAMI UNIVERSITY

University Libraries

Primary Unit: Center for Digital Scholarship

Primary Contact:
John Millard
Head, Center for Digital Scholarship
513-529-6789
millarj@miamioh.edu

PROGRAM OVERVIEW

Mission/description: We want to serve as a collaborative partner with faculty, students, and staff by providing infrastructure and expertise to support open access journals with or without peer review.

Year publishing activities began: 2011

Organization: centralized library publishing unit/department

Staff in support of publishing activities (FTE): library staff (0.5); undergraduate students (1)

PUBLISHING ACTIVITIES

Types of publications: faculty-driven journals (1); student-driven journals (1)

Media formats: text

Disciplinary specialties: computer science and engineering; psychology

Percentage of journals that are peer reviewed: 100

Campus partners: campus departments or programs; individual faculty; graduate students; undergraduate students

Publishing platform(s): OJS/OCS/OMP

Digital preservation strategy: in-house

Additional services: cataloging; metadata; author copyright advisory; digitization

MOUNT SAINT VINCENT UNIVERSITY

Mount Saint Vincent University Library

Primary Unit: Archives and Scholarly Communication
ojs@msvu.ca

Primary Contact:
Roger Gillis
Scholarly Communications and Archives Librarian
902-457-6401
roger.gillis@gmail.com

Website: journals.msvu.ca

PROGRAM OVERVIEW

Mission/description: Journals at the Mount is a hosting service provided by the Mount Saint Vincent University Library for the Mount community and/or affiliated partners. The service employs Open Journal Systems (OJS) as a the hosting platform for scholarly journals and includes training, support, and guidance for the development of new and existing publications of the Mount community.

Year publishing activities began: 2010

Organization: centralized library publishing unit/department

Staff in support of publishing activities (FTE): library staff (0.25)

Funding sources (%): library operating budget (70); other (30)

PUBLISHING ACTIVITIES

Types of publications: faculty-driven journals (2); ETDs (30); undergraduate capstone/honors theses (2)

Media formats: text; images

Disciplinary specialties: women's/gender studies; adult education

Top publications: *Atlantis: Critical Studies in Gender, Culture & Social Justice* (journal); *Canadian Journal for the Study of Adult Education* (journal)

Percentage of journals that are peer reviewed: 100

Campus partners: campus departments or programs; individual faculty; graduate students

Other partners: Canadian Association for the Study of Adult Education; Public Knowledge Project

Publishing platform(s): OJS/OCS/OMP

Digital preservation strategy: digital preservation services under discussion

Additional services: graphic design (print or web); marketing; outreach; training; analytics; cataloging; metadata; ISSN registration; DOI assignment/allocation of identifiers; contract/license preparation; author copyright advisory; hosting of supplemental content

Plans for expansion/future directions: Digitizing back issues, developing student journals, and discussing with faculty the development of new journals/migrating existing journals to the OJS platform.

NORTHEASTERN UNIVERSITY

University Libraries

Primary Unit: Scholarly Communication

Primary Contact:
Hillary Corbett
Scholarly Communication Librarian
617-373-2352
h.corbett@neu.edu

PROGRAM OVERVIEW

Mission/description: The University Libraries offer a growing suite of publishing services in response to the needs of faculty, students, and staff. The Libraries provide an online platform for journal publishing and the opportunity to produce innovative online collections and e-books through its digital repository service. Through the repository service, the Libraries also provide open access to the university's electronic theses and dissertations, scholarly research output, and university-produced objects.

Year publishing activities began: 2006

Organization: services are distributed across library units/departments

Staff in support of publishing activities (FTE): library staff (0.25)

Funding sources (%): library operating budget (100)

PUBLISHING ACTIVITIES

Types of publications: faculty-driven journals (2); student-driven journals (2); monographs (2); technical/research reports (1); ETDs (390); undergraduate capstone/honors theses (19)

Media formats: text; images

Percentage of journals that are peer reviewed: 100

Campus partners: campus departments or programs; individual faculty; graduate students; undergraduate students

Publishing platform(s): bepress (Digital Commons); Fedora; Omeka; Issuu

Digital preservation strategy: in-house; digital preservation services under discussion

Additional services: graphic design (print or web); typesetting; copy-editing; outreach; training; metadata; compiling indexes and/or TOCs; notification of A&I sources; DOI assignment/allocation of identifiers; dataset management; author copyright advisory; other author advisory; digitization; hosting of supplemental content

Plans for expansion/future directions: Working to expand the capabilities of our digital repository, in response to users' needs for space that can accommodate new kinds of projects; bringing another faculty journal online in the coming year.

NORTHWESTERN UNIVERSITY

Northwestern University Library

Library Publishing Coalition

Primary Unit: Center for Scholarly Communication and Digital Curation
cscdc@northwestern.edu

Primary Contact:
Claire Stewart
Head, Digital Collections and Scholarly Communication Services
847-467-1437
claire-stewart@northwestern.edu

Website: cscdc.northwestern.edu

Social media: @NU_CSCDC

PROGRAM OVERVIEW

Mission/description: We are engaged in planning activities to identify tools and support models that enable distributed, preservable publishing projects across the entire University. In initial phases, we anticipate the emphasis will be heavier on non-traditional products, transitioning to open theses, open journals, and open books as the key stakeholders, including our Press, move into closer technical and mission alignment.

Year publishing activities began: 2012

Organization: services are distributed across library units/departments

Staff in support of publishing activities (FTE): library staff (0.5)

Funding sources (%): library operating budget (94); charitable contributions/ Friends of the Library organizations (3); grants (3)

PUBLISHING ACTIVITIES

Types of publications: databases (1); scholarly websites that are heavily content-driven

Media formats: text; images

Disciplinary specialties: classics; history

Top publications: *Classicizing Chicago* (digital collection)

Campus partners: campus departments or programs; individual faculty

Publishing platform(s): Fedora; WordPress; Drupal

Digital preservation strategy: DuraCloud/DSpace; DPN; in-house; digital preservation services under discussion

Additional services: graphic design (print or web); training; metadata; dataset management; author copyright advisory; digitization; hosting of supplemental content

Additional information: We are working in many areas that blur into "library publishing," so it is sometimes hard to isolate the people, tasks, and funding that contribute to library publishing services. It is an area that we see as a growing component of our library's scholarly and digital programs. The fact that the University Press also reports to the Dean of Libraries opens up avenues for fruitful discussion, but to date the Press' publishing is quite separate from the Library's.

Plans for expansion/future directions: Developing a consulting service for faculty seeking to establish new publications and engaging in conversations with partners on campus around a shared investment in a cloud-based WordPress service, with plans to build and extend custom plugins for publishing projects and to integrate CMS-based publishing projects with the library's digital repository; exploring possible collaborations with the University Press, especially related to policy and infrastructure.

OBERLIN COLLEGE

Oberlin College Library

Primary Unit: Oberlin College Library
alan.boyd@oberlin.edu

Primary Contact:
Alan Boyd
Associate Director of Libraries
440-775-5015
alan.boyd@oberlin.edu

PROGRAM OVERVIEW

Mission/description: Publish all current and retrospective honors papers and master's theses with concurrence of the faculty department.

Year publishing activities began: 2008

Organization: centralized library publishing unit/department

Staff in support of publishing activities (FTE): library staff (0.15)

Funding sources (%): library operating budget (100)

PUBLISHING ACTIVITIES

Types of publications: undergraduate capstone/honors theses (150)

Media formats: text; images; multimedia/interactive content

Campus partners: campus departments or programs

Publishing platform(s): OhioLINK ETD Center

Digital preservation strategy: no digital preservation services provided

Additional services: outreach; cataloging; metadata; author copyright advisory; digitization; hosting of supplemental content

OHIO STATE UNIVERSITY

University Libraries

Library Publishing Coalition

Primary Unit: Digital Content Services
schlosser.40@osu.edu

Primary Contact:
Melanie Schlosser
Digital Publishing Librarian
614-688-5877
schlosser.40@osu.edu

Website: library.osu.edu/projects-initiatives/knowledge-bank

PROGRAM OVERVIEW

Mission/description: Our mission is to engage with partners across the university to increase the amount, value, and impact of OSU-produced digital content.

Year publishing activities began: 2004

Organization: centralized library publishing unit/department

Staff in support of publishing activities (FTE): library staff (3.5); undergraduate students (0.25)

Funding sources (%): library operating budget (100)

PUBLISHING ACTIVITIES

Types of publications: faculty-driven journals (4); student-driven journals (2); journals produced under contract/MOU for external groups (1); monographs (4); technical/research reports (17); student conference papers and proceedings (26); newsletters (6); undergraduate capstone/honors theses (447); conference and event lectures and presentations (727); graduate student culminating papers and projects (43); graduate student research forum papers and symposia posters (44); undergraduate research forum presentations and posters (40)

Media formats: text; images; audio; video; data

Percentage of journals that are peer reviewed: 100

Campus partners: campus departments or programs; individual faculty; graduate students; undergraduate students

Other partners: Society for Disability Studies; The Ohio Academy of Science

Publishing platform(s): DSpace; OJS/OCS/OMP

Digital preservation strategy: digital preservation services under discussion

Additional services: graphic design (print or web); typesetting; training; analytics; cataloging; metadata; compiling indexes and/or TOCs; notification of A&I sources; DOI assignment/allocation of identifiers; contract/license preparation; author copyright advisory; digitization; hosting of supplemental content; consulting and educational programming

Additional information: Although an ETD program is considered library publishing for this survey, we did not include ETDs. Our dissertations and theses are submitted by students to the OhioLINK consortial ETD database. Since autumn term 2002, dissertations have been produced by the student in electronic format and submitted to the OhioLINK ETD Center. Beginning calendar 2009, all master's theses have been produced by the student in electronic format and submitted to the OhioLINK ETD Center. We do not host our dissertations and theses separately.

Plans for expansion/future directions: Formalizing policies and procedures, recruiting new publishing partners, and adding new services.

HIGHLIGHTED PUBLICATION

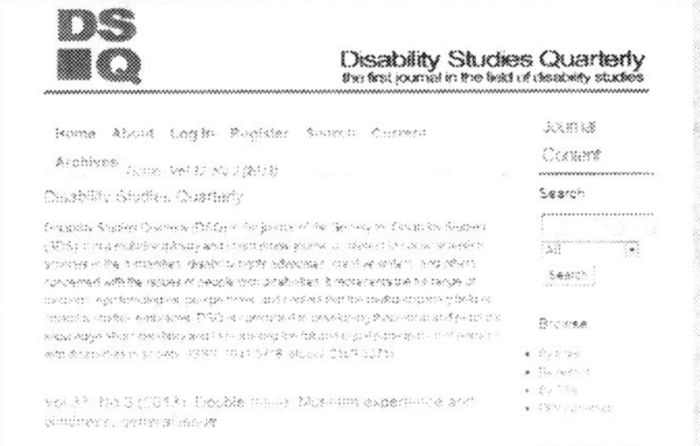

DS
Q
Disability Studies Quarterly
the first journal in the field of disability studies

Disability Studies Quarterly, the journal of the Society for Disability Studies, is a multidisciplinary, international publication that covers all aspects of disability studies.

dsq-sds.org

OREGON STATE UNIVERSITY

Oregon State University Libraries and Press

Primary Unit: Center for Digital Scholarship and Services

Primary Contact:
Michael Boock
Head of the Center for Digital Scholarship and Services
541-737-9155
michael.boock@oregonstate.edu

Website: cdss.library.oregonstate.edu

PROGRAM OVERVIEW

Mission/description: Oregon State University Libraries' publishing activities are primarily focused on the dissemination of scholarship produced by OSU faculty and students. This is achieved largely through the institutional repository ScholarsArchive@OSU, which includes previously unpublished material such as electronic theses and dissertations, agricultural extension reports, and faculty datasets. OSU Libraries also hosts open access journals that include articles by OSU faculty. The Libraries' Center for Digital Scholarship and Services digitizes selected out-of-print OSU Press publications, and provides open access to excerpts from Press books and supplementary materials such as maps and datasets. Other publishing activities involve the development of online resources that present and interpret unique holdings of OSU Libraries. Examples include extensive documentary histories and online exhibits on the Linus Pauling Papers and related archival collections in the History of Science and other areas. OSU Libraries has also developed digital resources in conjunction with books published by the OSU Press. Examples include a mobile application for touring historic buildings that is based on a book about Portland architecture, and a website that supports nature exploration related to a children's book published by the Press.

Year publishing activities began: 2006

Organization: services are distributed across library units/departments

Staff in support of publishing activities (FTE): library staff (3)

Funding sources (%): library operating budget (100)

PUBLISHING ACTIVITIES

Types of publications: faculty-driven journals (2); journals produced under contract/MOU for external groups (1); technical/research reports (50); faculty conference papers and proceedings (965); student conference papers and proceedings (125); ETDs (670); undergraduate capstone/honors theses (35); datasets

Media formats: text; images; audio; video; data

Disciplinary specialties: forestry; agriculture; history of science; water studies

Top publications: *Growing Your Own* (technical report); *Forest Phytophthoras* (journal); *International Institute for Fisheries Economics and Trade Conference Proceedings* (conference proceedings); *Journal of the Transportation Research Forum* (journal); *Reducing Fire Risk on Your Forest Property* (technical report)

Percentage of journals that are peer reviewed: 100

Campus partners: campus departments or programs; individual faculty; graduate students; undergraduate students

Other partners: Transportation Research Forum; International Institute for Fisheries Economics and Trade; Western Dry Kiln Association; Oregon Institute for Natural Resources

Publishing platform(s): CONTENTdm; DSpace; Fedora; OJS/OCS/OMP; WordPress; Omeka

Digital preservation strategy: Archive-It; LOCKSS; MetaArchive

Additional services: graphic design (print or web); training; analytics; cataloging; metadata; ISSN registration; DOI assignment/allocation of identifiers; dataset management; author copyright advisory; digitization; hosting of supplemental content; audio/video streaming

HIGHLIGHTED PUBLICATION

For more than a century, Oregon State University's Extension Service and Agricultural Experiment Station publications have covered everything from winemaking techniques to marine economics.

ir.library.oregonstate.edu/xmlui/handle/1957/3904

Additional Information: It should be noted that while the OSU Press is part of the OSU Libraries organization, the Press' publishing program, which results in the publication of approximately twenty-five books per year on the Pacific Northwest, has mostly operated independently from the Libraries' publishing activities. Therefore, the descriptions of "library publishing" have not included the Press' current print publishing output. In the future, the publishing programs of the Libraries and Press will be increasingly integrated.

Plans for expansion/future directions: Our plans for the future largely focus on open access student journals, digital humanities, and open textbooks. Student journals will publish research from OSU undergraduate and graduate students, as well as students from around the world in specific disciplines. Digital humanities projects will incorporate platforms that emphasize multimedia elements in presenting scholarship by OSU faculty. Open textbooks will involve a new partnership between the OSU Libraries and Press and the OSU Extended Campus Open Educational Resources unit to support development of open textbooks by OSU faculty. The OSU Libraries' Gray Family Chair for Innovative Library Services will focus on digital publishing for at least the next three years, with a new incumbent providing vision and direction for innovation and sustainability in digital publishing.

PACIFIC UNIVERSITY

Pacific University Libraries

Contributing Institution
Library Publishing Coalition

Primary Unit: Local Collections and Publication Services

Primary Contact:
Isaac Gilman
Scholarly Communications and Research Services Librarian
503-352-7209
gilmani@pacificu.edu

Website: www.pacificu.edu/library/services/lcps/index.cfm

PROGRAM OVERVIEW

Mission/description: Pacific University Libraries' publishing services exist to disseminate diverse and significant scholarly and creative work, regardless of a work's economic potential. Through flexible open access publishing models and author services, Pacific University Libraries will contribute to the discovery of new ideas (from scholars within and outside the Pacific community) and to the sustainability of the publishing system.

Year publishing activities began: 2010

Organization: centralized library publishing unit/department

Staff in support of publishing activities (FTE): library staff (0.75)

Funding sources (%): library operating budget (100)

PUBLISHING ACTIVITIES

Types of publications: faculty-driven journals (4); monographs (1); ETDs (105)

Media formats: text; images; audio

Disciplinary specialties: health care; philosophy; undergraduate research; librarianship

Top publications: *Essays in Philosophy* (journal); *Journal of Librarianship and Scholarly Communication* (journal); *Health & Interprofessional Practice* (journal)

Percentage of journals that are peer reviewed: 100

Campus partners: campus departments or programs; individual faculty; undergraduate students

Publishing platform(s): bepress (Digital Commons)

Digital preservation strategy: digital preservation services under discussion

Additional services: typesetting; copy-editing; training; analytics; notification of A&I sources; ISSN registration; DOI assignment/allocation of identifiers; author copyright advisory; digitization

PENNSYLVANIA STATE UNIVERSITY

Penn State University Libraries

Primary Unit: Publishing & Curation Services

Primary Contact:
Linda Friend
Head, Scholarly Publishing Services
814-865-0673
lxf5@psu.edu

Website: www.libraries.psu.edu/psul/pubcur.html

PROGRAM OVERVIEW

Mission/description: Our mission is to provide authors and researchers with consultation on publishing options and practical, alternative ways for Penn State faculty and students to publish and disseminate research in many formats. In addition, we provide assistance to scholarly journals and societies in disseminating their publications and proceedings electronically. We subscribe to the principles of open access to research information. Doctoral dissertations and master's theses for most academic programs are submitted digitally and are disseminated through the Libraries, and there is an active program of collecting and making student research available. The three primary research journals in the field of Pennsylvania history are part of our digitized collections. We are currently investigating the need and feasibility of offering an enhanced program of tiered publishing services, particularly for research journals, data, conference proceedings, and student-initiated work.

Year publishing activities began: 2000

Organization: centralized library publishing unit/department. Some operations and publishing workflow responsibilities are distributed among several library units/departments including technology support, cataloging, preservation, etc.

Staff in support of publishing activities (FTE): library staff (2.5); graduate students (0.5)

Funding sources (%): library operating budget (99); sales revenue (1)

PUBLISHING ACTIVITIES

Types of publications: faculty-driven journals (1); journals produced under contract/MOU for external groups (3); faculty conference papers and proceedings (1); ETDs (1155); undergraduate capstone/honors theses (575); graduate student research exhibition posters; undergraduate student research exhibition posters

Media formats: text; images; audio; video; data

Disciplinary specialties: Pennsylvania history and culture

Top publications: *Pennsylvania History Journal* (journal); *Pennsylvania Magazine of History and Biography* (magazine); *Western Pennsylvania History* (journal); *WEPAN Conference Proceedings* (conference proceedings)

Percentage of journals that are peer reviewed: 75

Campus partners: campus departments or programs; individual faculty; graduate students; undergraduate students

Other partners: Women in Engineering ProActive Network (WEPAN); Historical Society of Pennsylvania; Heinz History Center; Pennsylvania History Association

Publishing platform(s): CONTENTdm; OJS/OCS/OMP; WordPress

Digital preservation strategy: digital preservation services under discussion; digital preservation special team is currently working on a long range plan.

Additional services: marketing; outreach; metadata; dataset management; other author advisory; digitization; hosting of supplemental content

Plans for expansion/future directions: Redescribing the program, with expansion of services in the near future.

HIGHLIGHTED PUBLICATION

Western Pennsylvania History from the Heinz History Center is a colorful regional quarterly of interest to scholars and history buffs alike.

ojs.libraries.psu.edu/index.php/wph

PEPPERDINE UNIVERSITY

Pepperdine University Libraries

Primary Unit: Office of the Dean of Libraries

Primary Contact:
Mark Roosa
Dean of Libraries
310-506-4252
mark.roosa@pepperdine.edu

Website: digitalcommons.pepperdine.edu

PROGRAM OVERVIEW

Mission/description: The Pepperdine Libraries provide a global gateway to knowledge, serving the diverse and changing needs of our learning community through personalized service at our campus locations and rich computer-based resources. At the academic heart of our educational environment, our libraries are sanctuaries for study, learning, and research, encouraging discovery, contemplation, social discourse, and creative expression. As the information universe continues to evolve, our goal is to remain responsive to users' needs by providing seamless access to both print and digital resources essential for learning, teaching, and research. The libraries, through Digital Commons@Pepperdine, offer a wide array of digital publications that are openly available for study, research, and learning.

Year publishing activities began: 2010

Organization: centralized library publishing unit/department

Staff in support of publishing activities (FTE): library staff (1)

Funding sources (%): library operating budget (100)

PUBLISHING ACTIVITIES

Types of publications: faculty-driven journals (1); student-driven journals (7); journals produced under contract/MOU for external groups (1); newsletters (3); ETDs (108); undergraduate capstone/honors theses (11)

Media formats: text; images; audio; data

Disciplinary specialties: religion; business; public policy; psychology; law

Top publications: *Pepperdine Law Review* (journal); *Leaven* (journal); *Pepperdine Dispute Resolution Law Journal* (journal); *The Journal of Business, Entrepreneurship and the Law* (journal); *Journal of the National Association of Administrative Law Judiciary* (journal)

Percentage of journals that are peer reviewed: 100

Campus partners: campus departments or programs; individual faculty; graduate students; undergraduate students

Publishing platform(s): bepress (Digital Commons); CONTENTdm

Digital preservation strategy: LOCKSS; Portico; Preservica; in-house

Additional services: marketing; outreach; training; cataloging; metadata; dataset management; digitization; audio/video streaming

Plans for expansion/future directions: Publishing additional undergraduate research; creating a line of monographic publications; publishing rich media content (e.g., video presentations); implementing an enterprise digital preservation solution; identifying new ways of participating in the editorial processes generally associated with publishing.

PORTLAND STATE UNIVERSITY

Portland State University Library

Primary Unit: Digital Initiatives

Primary Contact:
Sarah Beasley
Scholarly Communication Coordinator
503-725-3688
bvsb@pdx.edu

PROGRAM OVERVIEW

Mission/description: Portland State University (PSU) Library provides the infrastructure and a suite of services to offer a publishing platform that facilitates open access distribution; enhanced web search engine discovery through standards-based metadata and file formatting; permanent URLs; file formatting and format migration; copyright advisory for authors; and outreach for and promotion of PSU faculty or PSU departmentally sponsored content.

Year publishing activities began: 2010

Organization: services are distributed across library units/departments

Staff in support of publishing activities (FTE): library staff (2.5)

Funding sources (%): library operating budget (100)

PUBLISHING ACTIVITIES

Types of publications: student-driven journals (1); monographs (1); student conference papers and proceedings (51); ETDs (2000); undergraduate capstone/ honors theses (26)

Media formats: text; images; audio; video; data

Disciplinary specialties: physics; environmental sciences; engineering and computer science; urban studies and planning; education

Percentage of journals that are peer reviewed: 0

Campus partners: campus departments or programs; individual faculty

Publishing platform(s): bepress (Digital Commons)

Digital preservation strategy: in-house

Additional services: marketing; outreach; analytics; cataloging; metadata; dataset management; peer review management; author copyright advisory; digitization; hosting of supplemental content

Plans for expansion/future directions: Hosting journals, archiving monographs, and producing open access textbooks.

PURDUE UNIVERSITY

Purdue University Libraries

Primary Unit: Purdue Scholarly Publishing Services

Primary Contact:
Charles Watkinson
Head, Scholarly Publishing Services
765-494-8251
ctwatkin@purdue.edu

Website: www.lib.purdue.edu/publishing

Social media: @PublishPurdue

PROGRAM OVERVIEW

Mission/description: Purdue Scholarly Publishing Services focuses on supporting the publication efforts of various centers and departments within the Purdue system. The primary publishing platform used is Purdue e-Pubs (www.purdue.edu/epubs), and the majority of products created are openly accessible, free-of-charge, to readers. Open access is made possible by the financial support of partners, foundations, and Purdue University Libraries. Major initiatives include the production of the *Journal of Purdue Undergraduate Research*, the publication of technical reports on behalf of the Joint Transportation Research Program (JTRP), and the project management of HABRI Central, a major bibliographic reference database for researchers in the area of human-animal bond studies, produced in partnership with the Purdue College of Veterinary Medicine. Purdue Scholarly Publishing Services and Purdue University Press, which publishes more formal books and journals, together constitute the publishing division of Purdue Libraries. Our diverse publishing activities are supported by a single group of staff members with assistance from undergraduate and graduate students. By harnessing the skills of both librarians and publishers, and leveraging a common infrastructure, we believe we can better serve the needs of scholars in the digital age and enhance the impact of Purdue scholarship by developing information products aligned with the University's strengths.

Year publishing activities began: 2006

Organization: centralized library publishing unit/department

Staff in support of publishing activities (FTE): library staff (4.5); graduate students (0.75); undergraduate students (2.5)

Funding sources (%): library operating budget (50); non-library campus budget (10); grants (15); sales revenue (3); licensing (3); charge backs (19)

PUBLISHING ACTIVITIES

Types of publications: faculty-driven journals (11); student-driven journals (3); monographs (1); textbooks; (3); technical/research reports (28); faculty conference papers and proceedings (5); ETDs (1063); undergraduate capstone/honors theses (5); HABRI Central (an information hub for human-animal bond studies built on the HUBzero platform for scientific collaboration); the Data Curation Profiles Directory

Media formats: text; images; audio; video; data; multimedia/interactive content.

Disciplinary specialties: engineering (civil engineering); education (STEM); library and information science; public policy; comparative literature

Top publications: *Joint Transportation Research Program Technical Reports* (technical reports); *JPUR: Journal of Purdue Undergraduate Research* (journal); HABRI Central (website); *CLCWeb: Comparative Literature and Culture* (journal); *Interdisciplinary Journal of Problem-based Learning* (journal)

Percentage of journals that are peer reviewed: 100

Campus partners: campus departments or programs; individual faculty; graduate students; undergraduate students

Other partners: Indiana Department of Transportation (INDOT); HABRI Foundation; Charleston Conference/Against the Grain Press; International Association of Scientific and Technological University Libraries (IATUL)

HIGHLIGHTED PUBLICATION

The *Journal of Purdue Undergraduate Research (JPUR)* has been established to publish outstanding research papers written by Purdue undergraduates from all disciplines who have completed faculty-mentored research projects.

docs.lib.purdue.edu/jpur

Publishing platform(s): bepress (Digital Commons); HUBzero for HABRI Central

Digital preservation strategy: CLOCKSS and Portico for most important journals; MetaArchive for HABRI Central

Additional services: graphic design (print or web); typesetting; copy-editing; marketing; outreach; training; analytics; cataloging; metadata; compiling indexes and/or TOCs; notification of A&I sources; ISSN registration; DOI assignment/ allocation of identifiers; open URL support; dataset management; peer review management; business model development; budget preparation; contract/license preparation; author copyright advisory; other author advisory; digitization; hosting of supplemental content; audio/video streaming; developmental editing; project management

Additional information: Data publication is handled by the Purdue University Research Repository (PURR), which is a collaborative project of the Libraries, Information Technology at Purdue (ITaP), and the Office of the Vice President for Research. We have classified all open access journals as being products of Scholarly Publishing Services because of the types of workflow adopted, but five of these use the Purdue University Press imprint.

Plans for expansion/future directions: Working to expand the number of centers and departments we serve on campus, particularly in the area of conference proceedings and technical reports; creating better linkages between publications and materials in Purdue's data and archival repositories; developing better capacity to handle multimedia and "new form" publications; developing a clearer sustainability plan across the Libraries Publishing Division that balances earned revenue with internal support.

ROCHESTER INSTITUTE OF TECHNOLOGY

The Wallace Center

Primary Unit: Scholarly Publishing Studio

Primary Contact:
Nick Paulus
Manager of Scholarly Publishing
585-475-7934
njpwml@rit.edu

Website: wallacecenter.rit.edu/scholarly-publishing-studio

PROGRAM OVERVIEW

Mission/description: We connect stakeholders' scholarship efforts with our comprehensive publishing services, ensuring that faculty and student research is made available to readers faster and disseminated in a way that meets their academic objectives. Our approach is collaborative. We offer help with design and layout, copy-editing outsourcing, open access publishing, and pre-publishing consultation. At SPS, we are committed to advancing the dissemination of scholarship.

Organization: centralized library publishing unit/department

PUBLISHING ACTIVITIES

Campus partners: campus departments or programs; individual faculty

Publishing platform(s): bepress (Digital Commons); OJS/OCS/OMP

Additional services: graphic design (print or web); copy-editing

RUTGERS, THE STATE UNIVERSITY OF NEW JERSEY

Rutgers University Libraries

Contributing Institution Library Publishing Coalition

Primary Unit: Scholarly Communication Center

Primary Contact:
Rhonda Marker
RUcore Collection Manager/Head, Scholarly Communications Center
848-932-5923
rmarker@rutgers.edu

Website: rucore.libraries.rutgers.edu/services

PROGRAM OVERVIEW

Mission/description: The goal of the Rutgers University Community Repository is to advance research and learning at Rutgers, to foster interdisciplinary collaboration, and to contribute to the development of new knowledge through the archiving, preservation, and presentation of digital resources. Original research products and papers of the faculty and administrators and the unique resources of the libraries will be permanently preserved and made accessible with tools developed to facilitate and encourage their continued use.

Year publishing activities began: 2002

Organization: services are distributed across library units/departments

Staff in support of publishing activities (FTE): library staff (16); graduate students (3)

Funding sources (%): library materials budget (80); grants (20)

PUBLISHING ACTIVITIES

Types of publications: faculty-driven journals (5); journals produced under contract/MOU for external groups (1); faculty conference papers and proceedings (3); databases (1); ETDs (827); research/interview videos (200); a new scholarly communication form, the published video analytic, currently in use in our NSF-funded mathematics education collection, the Video Mosaic (www.videomosaic.org)

Media formats: text; video; data; multimedia/interactive content

Disciplinary specialties: mathematics education; psychology; jazz music; New Jersey history; classical studies

Top publications: Video Mosaic Collaborative (website); *Pragmatic Case Studies in Psychotherapy* (journal); *Journal of Jazz Studies* (journal); *Journal of Rutgers University Libraries* (journal); *New Jersey History* (journal)

Percentage of journals that are peer reviewed: 100

Campus partners: individual faculty; graduate students

Publishing platform(s): Fedora; OJS/OCS/OMP; locally developed software

Digital preservation strategy: in-house

Additional services: graphic design (print or web); training; analytics; cataloging; metadata; notification of A&I sources; DOI assignment/allocation of identifiers; contract/license preparation; author copyright advisory; audio/video streaming

Additional Information: Our focus is on the unique scholarship and resources of Rutgers University and on the research and education needs of our community. We develop new tools and services, including new modes of scholarly communication, in response to faculty and student needs, often through collaboration in research grants.

Plans for expansion/future directions: Expanding our publishing of original research and scholarship, with a particular focus on research data and digital video, including video of conferences and lectures held in the Alexander Library; exploring the publishing of undergraduate research in open access journals and new modes of scholarly communication, particularly in the humanities and social sciences.

SIMON FRASER UNIVERSITY

Simon Fraser University Library

1. THESES/DISSERTATIONS

Primary Unit: Thesis Office

Primary Contact:
Nicole White
Head, Research Commons
778-782-3268
ngjertse@sfu.ca

Website: www.lib.sfu.ca/help/writing/thesis

PROGRAM OVERVIEW

Mission/description: Responsible for accepting formatted theses and dissertations, and depositing them in the Library's institutional repository, Summit. Summit also acts as a publication platform for University authors (e.g., conference papers, technical reports). Conforms to OAI-PMH.

Year publishing activities began: 2004

Organization: services are distributed across library units/departments

Staff in support of publishing activities (FTE): library staff (2)

Funding sources (%): library operating budget (100%)

PUBLISHING ACTIVITIES

Types of publications: technical/research reports (39); faculty conference papers and proceedings (40); ETDs (563); undergraduate capstone/honors theses (1)

Media formats: text; images; audio; video; data; multimedia/interactive content

Campus partners: campus departments and programs; individual faculty; graduate students,

Publishing platform(s): Drupal

Digital preservation strategy: Archivematica, in-house. Work is underway to allow Archivematica to store AIP's (archival information packages) in a COPPUL Private LOCKSS Network.

Additional services: copy-editing; training; analytics; compiling indexes and/or TOCs; author copyright advisory; hosting of supplemental content

Plans for expansion/future directions: Working toward becoming a Trusted Digital Repository. Moved to exclusively digital thesis submission.

2. SCHOLARLY JOURNALS AND CONFERENCES

Primary Unit: Public Knowledge Project Publishing Services (PKP|PS)
pkp-hosting@sfu.ca

Primary Contact:
Brian Owen
Associate University Librarian/PKP Managing Director
778-782-7095
brian_owen@sfu.ca

Website: http://www.lib.sfu.ca/collections/scholarly-publishing; https://pkpservices.sfu.ca

PROGRAM OVERVIEW

Mission/description: Provide online hosting and related technical support at no charge for scholarly journals and conferences that have a significant SFU faculty connection (e.g., a Managing Editor) or to support SFU-based teaching and research initiatives.

Year publishing activities began: 2005

Organization: The SFU Library provides the administrative and technical home for PKP and its related activities, such as PKP Publishing Services. In return, PKP|PS provides the technical expertise and infrastructure support for the SFU Library's scholarly communication services. PKP|PS staff work closely with the Library's liaison librarians.

Staff in support of publishing activities (FTE): library staff (.2)

Funding sources (%): library operating budget (25); PKP|PS in-kind (75)

PUBLISHING ACTIVITIES

Types of publications: faculty-driven and graduate student journals (10); scholarly conferences (2)

Media formats: text; images; audio; video; data; multimedia/interactive content

Other partners: SFU's Canadian Centre for Studies in Publishing

Publishing platform(s): OJS/OCS

Digital preservation strategy: COPPUL; LOCKSS

Additional services: digitization; software customization/development

Additional Information: PKP Publishing Services is not a typical library publishing operation. By virtue of being the developers of OJS and other PKP software, we are able to offer technical support that may not be feasible for other library publishing services.

Plans for expansion/future directions: Hosting and related support for Open Monograph Press (OMP).

STATE UNIVERSITY OF NEW YORK AT BUFFALO

E. H. Butler Library

Primary Unit: Scholarly Communication Librarian

Primary Contact:
Marc D. Bayer
Scholarly Communication Librarian
716-878-6305
bayermd@buffalostate.edu

Website: digitalcommons.buffalostate.edu/submit_research.html

PROGRAM OVERVIEW

Mission/description: The E. H. Butler Library publishes monographs and periodicals that feature the research, applied, and artistic works of the Buffalo State community. In addition to a print publishing program, the library administers the campus institutional repository.

Year publishing activities began: 2009

Organization: centralized library publishing unit/department

Staff in support of publishing activities (FTE): library staff (1)

PUBLISHING ACTIVITIES

Campus partners: campus departments or programs; individual faculty; graduate students; undergraduate students

Other partners: Monroe Fordham Regional History Center of SUNY Buffalo State

Plans for expansion/future directions: Including more student research.

STATE UNIVERSITY OF NEW YORK AT GENESEO

Milne Library

Primary Unit: Technical Services
milne@geneseo.edu

Primary Contact:
Allison Brown
Editor and Production Manager
585-245-6020
browna@geneseo.edu

Website: publishing.geneseo.edu

PROGRAM OVERVIEW

Mission/description: The mission of Milne Library publishing services is based on a core value of libraries: knowledge sharing and literacy are an essential public good. The goal of Milne publishing is to inspire authors and creators to share their works with a sustainable publishing model that rewards both authors and readers, libraries and learning. Milne publishing will help transform scholarly communications and library publishing.

Year publishing activities began: 2012

Organization: distributed across library units; Open SUNY Textbooks and individual journals are distributed among various institutions

Staff in support of publishing activities (FTE): library staff (3)

Funding sources (%): library operating budget (45); grants (55)

PUBLISHING ACTIVITIES

Types of publications: faculty-driven journals (2); student-driven journals (1); monographs (7); student conference papers and proceedings (1); newsletters (2); TEI digital humanities projects; OMEKA digital collections; best practices toolkits

Media formats: text; images.

Disciplinary specialties: education; library and information science; local history; humanities/liberal arts

Top publications: digitalthoreau.org (website); reprints and new monographs on Amazon.com; *Educational Change* (journal); reprints on Open Monograph Press; *Workflow Toolkit* (website)

Percentage of journals that are peer reviewed: 100

Campus partners: campus departments or programs; individual faculty; undergraduate students

Other partners: Thoreau Society; Thoreau Institute; Walden Woods Project; New York State Foundations of Education Association

Publishing platform(s): CONTENTdm; OJS/OCS/OMP; WordPress; Commons in a Box

Digital preservation strategy: no digital preservation services provided; server backup as appropriate

Additional services: graphic design (print or web); typesetting; copy-editing; marketing; outreach; training; analytics; cataloging; metadata; compiling indexes and/or TOCs; ISSN registration; DOI assignment/allocation of identifiers; open URL support; peer review management; business model development; budget preparation; contract/license preparation; author copyright advisory; other author advisory; digitization; hosting of supplemental content; audio/video streaming

Additional Information: *Library Publishing Toolkit* available at: www.publishingtoolkit.org. Developing interactives with video, multiple choice feedback, etc.

Plans for expansion/future directions: Expanding the use of Open Monograph Press for textbook, reprints, and new monograph publishing; developing network hosting and training models for Open Journal Systems and Open Monograph Press; expanding the role of digital scholarship publishing with social reading in Digital Thoreau and the use of Omeka.

SYRACUSE UNIVERSITY

Syracuse University Libraries

Primary Unit: Scholarly Communication

Primary Contact:
Yuan Li
Scholarly Communication Librarian
315-443-4247
yli115@syr.edu

Website: surface.syr.edu

PROGRAM OVERVIEW

Mission/description: To provide Syracuse University (SU) faculty with an alternative to commercial publishing venues, and to provide the campus community support for open access publishing models.

Year publishing activities began: 2010

Organization: services are distributed across library units/departments

Staff in support of publishing activities (FTE): library staff (3)

Funding sources (%): library materials budget (25); library operating budget (75)

PUBLISHING ACTIVITIES

Types of publications: faculty-driven journals (1); student-driven journals (2); monographs (4); technical/research reports (98); faculty conference papers and proceedings (48); student conference papers and proceedings (15); newsletters (290); ETDs (248); working papers; journal articles; images; video; and presentations

Media formats: text; video

Disciplinary specialties: law and commerce; public diplomacy; writing and rhetoric; disability and popular culture

Top publications: *Intertext* (journal)

Percentage of journals that are peer reviewed: 100

Campus partners: Syracuse University Press; campus departments or programs; individual faculty; graduate students; undergraduate students

Publishing platform(s): bepress (Digital Commons); OJS/OCS/OMP

Digital preservation strategy: APTrust; DPN; LOCKSS

Additional services: graphic design (print or web); typesetting; copy-editing; marketing; training; analytics; cataloging; metadata; ISSN registration; DOI assignment/allocation of identifiers; open URL support; peer review management; business model development; author copyright advisory; digitization; hosting of supplemental content; audio/video streaming

Plans for expansion/future directions: Launching a joint imprint (Syracuse Unbound) with Syracuse University Press and two new open access journals in the coming months; forming a new unit that brings together several units involved in digital scholarship activities, including digital publishing; formalizing a menu of publishing services for the campus community.

HIGHLIGHTED PUBLICATION

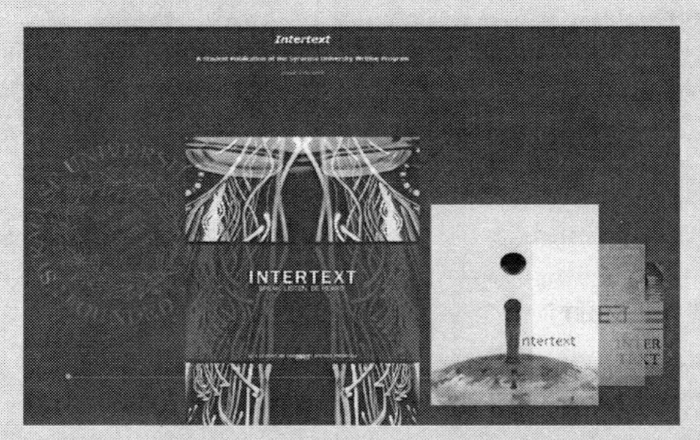

Intertext aims to represent the writing of Syracuse University students through publishing exemplary works submitted from any Writing Program undergraduate course.

wrt-intertext.syr.edu

TEMPLE UNIVERSITY

Temple University Libraries

Primary Unit: Digital Library Initiatives
diglib@temple.edu

Primary Contact:
Delphine Khanna
Head of Digital Library Initiatives
215-204-4768
delphine@temple.edu

Website: digital.library.temple.edu

PROGRAM OVERVIEW

Mission/description: The goal of our program is to provide free and open access to digital scholarship produced by Temple University students. Currently, we focus on the publishing of doctoral dissertations, master's theses, and the winning essays of the Temple University Library Prize for Undergraduate Research in general topics and in topics related to sustainability and the environment. In the future, we plan to greatly expand our publishing program to include scholarly journals and books.

Year publishing activities began: 2008

Organization: services are distributed across library units/departments

Staff in support of publishing activities (FTE): library staff (0.5)

Funding sources (%): library operating budget (100)

PUBLISHING ACTIVITIES

Types of publications: ETDs (400); winning essays for the Temple University Library Prize for Undergraduate Research (7)

Media formats: text; images; data

Disciplinary specialties: full range of academic subjects in ETDs

Top publications: "The Digitalization of Music Culture: A Case Study Examining the Musician/Listener Relationship with Digital Technology" (thesis); "Profitability Ratio Analysis for Professional Service Firms" (thesis); "Naskh Al-Qur'an: A Theological and Juridical Reconsideration of the Theory of Abrogation and Its Impact on Qur'anic Exegesis" (thesis); "Pcaob International Inspection and Audit Quality" (thesis); "Mother of God, Cease Sorrow!: The Significance of Movement in a Late Byzantine Icon" (thesis)

Campus partners: campus departments or programs

Publishing platform(s): CONTENTdm

Digital preservation strategy: in-house. digital preservation services under discussion; our CONTENTdm instance is hosted at OCLC and they have backup procedures. We are also now considering membership in HathiTrust.

Additional services: analytics; cataloging; metadata; hosting of supplemental content

Plans for expansion/future directions: Planning significant expansion of services, such as the inclusion of books and journals.

TEXAS TECH UNIVERSITY

Texas Tech University Libraries

Primary Unit: Digital Resources Library Unit

Primary Contact:
Christopher Starcher
Digital Services Librarian
806-834-0855
christopher.starcher@ttu.edu

PROGRAM OVERVIEW

Mission/description: To publish and archive the scholarship of Texas Tech University by its faculty, researchers, and students.

Year publishing activities began: 2005

Organization: centralized library publishing unit/department

Staff in support of publishing activities (FTE): library staff (2); graduate students (0.75); undergraduate students (0.5)

Funding sources (%): library operating budget (100)

PUBLISHING ACTIVITIES

Types of publications: textbooks (1); student conference papers and proceedings (1); ETDs (1200); undergraduate capstone/honors theses (50)

Media formats: text; images

Top publications: ETDs; honors theses

Campus partners: campus departments or programs; individual faculty; graduate students; undergraduate students

Publishing platform(s): DSpace

Digital preservation strategy: DuraCloud/DSpace; LOCKSS; Scholars Portal; in-house; digital preservation services under discussion. Everything is housed at the university data center and then backed up to an out-of-town remote storage facility.

Additional services: outreach; training; analytics; metadata; DOI assignment/ allocation of identifiers; author copyright advisory; other author advisory; digitization

THOMAS JEFFERSON UNIVERSITY

Scott Memorial Library

Primary Unit: Academic and Instructional Support & Resources

Primary Contact:
Dan Kipnis
Senior Education Services Librarian and Editor of Jefferson Digital Commons
215-503-2825
Dan.kipnis@jefferson.edu

PROGRAM OVERVIEW

Mission/description: To provide an open access institutional repository of the work being produced by the Jefferson community to a global audience.

Year publishing activities began: 2006

Organization: centralized library publishing unit/department

Staff in support of publishing activities (FTE): library staff (1.5)

Funding sources (%): library materials budget (100)

PUBLISHING ACTIVITIES

Types of publications: faculty-driven journals (3); student-driven journals (2); journals produced under contract/MOU for external groups (1); monographs (4); technical/research reports (3); student conference papers and proceedings (20); newsletters (4); ETDs (50); videos; grand round presentations; conference posters

Media formats: text; images; audio; video

Disciplinary specialties: historical psychiatry; internal medicine; population studies; integrative medicine

Top publications: *Jefferson Journal of Psychiatry* (journal); *The Medicine Forum* (journal); *On the Anatomy of the Breast* (monograph); *A Manual of Military Surgery* (monograph); *Legend and Lore: Jefferson Medical College* (monograph)

Percentage of journals that are peer reviewed: 100

Campus partners: campus departments or programs; individual faculty

Other partners: Special Library Association

Publishing platform(s): bepress (Digital Commons)

Digital preservation strategy: LOCKSS; in-house

Additional services: marketing; outreach; training; analytics; metadata; ISSN registration; author copyright advisory; digitization; hosting of supplemental content; audio/video streaming

Additional information: We are encouraged by our grass roots effort to get materials in our IR and to continue our publishing efforts.

Plans for expansion/future directions: Continuing to add journals, newsletters, and additional grey literature materials to our institutional repository.

TRINITY UNIVERSITY

Coates Library

Primary Unit: Discovery Services

Primary Contact:
Jane Costanza
Head of Discovery Services
210-999-7612
jcostanz@trinity.edu

Website: digitalcommons.trinity.edu

PROGRAM OVERVIEW

Mission/description: The Trinity University open access policy encourages faculty authors to retain non-commercial copyright for their scholarly publications and provides them with the means to negotiate those rights with their publishers. Additionally, open access facilitates the sharing of peer-reviewed research through Trinity's digital repository (Digital Commons @ Trinity), which provides broad, free access to a faculty author's scholarly work. The open access policy at Trinity depends for its effectiveness on faculty authors granting to the university permission to upload digital copies of their scholarly publications to Trinity's digital repository.

Year publishing activities began: 2009

Organization: centralized library publishing unit/department

Staff in support of publishing activities (FTE): library staff (1); undergraduate students (0.05)

Funding sources (%): library operating budget (100)

PUBLISHING ACTIVITIES

Types of publications: faculty-driven journals (1); journals produced under contract/MOU for external groups (1); monographs (4); textbooks; (1); technical/research reports (3); student conference papers and proceedings (36); undergraduate capstone/honors theses (22); administrative reports

Media formats: text; images; video; data

Disciplinary specialties: teacher education; anthropology; psychology; mathematics; biology

Top publications: "Cognitive Bias Modification: Past Perspectives, Current Findings, and Future Applications" (thesis); "Cognitive Bias Modification: Induced Interpretive Biases Affect Memory" (thesis); "A Survey of Psychologists' Attitudes Towards and Utilization of Exposure Therapy for PTSD" (thesis); "Islamophobia, Euro-Islam, Islamism and Post-Islamism: Changing Patterns of Secularism in Europe" (thesis); *Tipiti* (journal)

Percentage of journals that are peer reviewed: 50

Campus partners: campus departments or programs; individual faculty; undergraduate students

Other partners: Society for the Anthropology of Lowland South America

Publishing platform(s): bepress (Digital Commons)

Digital preservation strategy: CLOCKSS

Additional services: analytics; cataloging; metadata; open URL support; author copyright advisory

Additional information: We also support SelectedWorks.

Plans for expansion/future directions: Continuing to help faculty members understand the issues around the economics of scholarly publishing and the benefits of providing open access to their scholarly output.

TULANE UNIVERSITY

Howard-Tilton Memorial Library

Primary Unit: Digital Initiatives

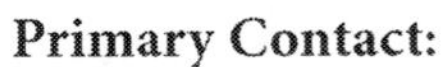

Primary Contact:
Jeff Rubin
Digital Initiatives and Publishing Coordinator
504-247-1832
jrubin6@tulane.edu

Website: library.tulane.edu/repository

PROGRAM OVERVIEW

Mission/description: Tulane University Journal Publishing is an open access journal publishing service that provides a web-based platform for scholarly and academic publishing to the Tulane community.

Year publishing activities began: 2012

Organization: centralized library publishing unit/department

Staff in support of publishing activities (FTE): library staff (1)

Funding sources (%): library operating budget (100)

PUBLISHING ACTIVITIES

Types of publications: faculty-driven journals (1); student-driven journals (2); journals produced under contract/MOU for external groups (1); ETDs (100)

Media formats: text; images; audio; video

Disciplinary specialties: zoology; botany; international affairs; literary

Top publications: *Tulane Studies in Zoology and Botany* (journal); *Tulane Review* (journal); *Tulane Journal of International Affairs* (journal); *Second Line: An Undergraduate Journal of Literary Conversation* (journal)

Percentage of journals that are peer reviewed: 100

Campus partners: campus departments or programs; individual faculty; graduate students; undergraduate students

Publishing platform(s): OJS/OCS/OMP

Digital preservation strategy: DPN; centralized storage and backup through Tulane Technology Services

Additional services: training; metadata; ISSN registration; author copyright advisory; other author advisory; hosting of supplemental content; audio/video streaming

UNIVERSITÉ DE MONTRÉAL

Université de Montréal Libraries

Primary Unit: Teaching, Learning and Research Support

Primary Contact:
Diane Sauvé
Director, Teaching, Learning and Research Support
514-343-6111 ext. 1817
diane.sauve@umontreal.ca

Website: www.bib.umontreal.ca/Papyrus

PROGRAM OVERVIEW

Mission/description: The Université de Montréal institutional repository, Papyrus, provides access to the university theses and dissertations, as well as to some publications and other forms of intellectual output from the university.

Year publishing activities began: 2003

Organization: centralized library publishing unit/department

Staff in support of publishing activities (FTE): library staff (1)

Funding sources (%): library operating budget (100)

PUBLISHING ACTIVITIES

Types of publications: ETDs (1200)

Media formats: text; images; audio; video

Campus partners: campus departments or programs

Publishing platform(s): DSpace

Digital preservation strategy: no digital preservation services provided

UNIVERSITY OF ALBERTA

University of Alberta Libraries

Primary Unit: Digital Initiatives

Primary Contact:
Leah Vanderjagt
Digital Repository Services Librarian
780-492-3851
leah.vanderjagt@ualberta.ca

Website: guides.library.ualberta.ca/oa

Social media: listed at www.library.ualberta.ca

PROGRAM OVERVIEW

Mission/description: The University of Alberta Libraries provides support to community members who want to publish in OA formats (e.g., providing journal hosting and institutional repository services).

Year publishing activities began: 2007

Staff in support of publishing activities (FTE): library staff (1.3); graduate students (0.1)

PUBLISHING ACTIVITIES

Types of publications: faculty-driven journals (24); student-driven journals (6); textbooks; (1); technical/research reports (30); faculty conference papers and proceedings (5); student conference papers and proceedings (3); ETDs (1200); undergraduate capstone/honors theses (8); videos

Media formats: text; images; audio; video; data; concept maps/modeling maps/visualizations; multimedia/interactive content

Disciplinary specialties: library and information studies; education; pharmaceutical sciences; sociology; environmental studies (particularly oil sands)

Top publications: *Canadian Journal of Sociology* (journal); *International Journal of Qualitative Methods* (journal); *Journal of Pharmacy & Pharmaceutical Sciences* (journal); *Evidence Based Library and Information Practice* (journal); *Canadian Review of Comparative Literature* (journal)

Percentage of journals that are peer reviewed: 75

Campus partners: campus departments or programs; individual faculty; graduate students; undergraduate students

Other partners: Public Knowledge Project; research teams/projects (e.g., Oil Sands Research and Information Network; Canadian Writing Research Collaboratory); local non-profit organizations (e.g., Edmonton Social Planning Council).

Publishing platform(s): Fedora; OJS/OCS/OMP; WordPress; locally developed software

Digital preservation strategy: Archive-It; Archivematica; CLOCKSS; COPPUL; HathiTrust; LOCKSS; Portico; in-house

Additional services: outreach; cataloging; metadata; DOI assignment/allocation of identifiers; open URL support; dataset management; author copyright advisory; other author advisory; digitization; hosting of supplemental content

Additional information: Funding for publishing services comes out of the library operations budget. However, we do not have a fixed breakdown. We do not charge users for our publishing services and only publish open access content.

Plans for expansion/future directions: Supporting the growth of our institutional repository and journal hosting services; facilitating the development of campus-wide scholarly publishing initiatives (e.g., establishing an open monograph publishing service, research data "publication" and curation), open educational resources (OER), etc.

UNIVERSITY OF ARIZONA

University of Arizona Libraries

Primary Unit: Scholarly Publishing and Data Management Team
repository@u.library.arizona.edu

Primary Contact:
Dan Lee
Director, Office of Copyright Management and Scholarly Communication
520-621-6433
leed@email.arizona.edu

Website: journals.uair.arizona.edu; arizona.openrepository.com/arizona

PROGRAM OVERVIEW

Mission/description: The Scholarly Publishing and Data Management Team provides tools, services, and expertise that enable the creation, distribution, and preservation of scholarly works and research data in support of the mission of the University of Arizona.

Year publishing activities began: 1994

Organization: centralized library publishing unit/department

Staff in support of publishing activities (FTE): library staff (3.25); graduate students (1); undergraduate students (2)

Funding sources (%): library operating budget (100)

PUBLISHING ACTIVITIES

Types of publications: faculty-driven journals (5); student-driven journals (3); journals produced under contract/MOU for external groups (5); monographs (1); technical/research reports (1391); student conference papers and proceedings (104); newsletters (1); ETDs (19660); undergraduate capstone/honors theses (1274); periodicals (8)

Media formats: text; images; audio; video; data

Disciplinary specialties: agriculture; life sciences; dendrochronology; archaeology; geosciences

Top publications: *Radiocarbon* (journal); *Journal of Ancient Egyptian Interconnections* (journal); ETDs; *Coyote Papers* (working papers); *Arizona Anthropologist* (journal)

Percentage of journals that are peer reviewed: 100

Campus partners: campus departments or programs; individual faculty; graduate students; undergraduate students

Other partners: International Society of Lymphology; Society for Range Management; Tree Ring Society

Publishing platform(s): CONTENTdm; DSpace; OJS/OCS/OMP; locally developed software

Digital preservation strategy: digital preservation services under discussion

Additional services: training; analytics; cataloging; metadata; notification of A&I sources; ISSN registration; DOI assignment/allocation of identifiers; dataset management; peer review management; contract/license preparation; author copyright advisory; other author advisory; digitization; hosting of supplemental content

Plans for expansion/future directions: Discussing collaborative efforts with the university press.

HIGHLIGHTED PUBLICATION

Radiocarbon is the main international journal of record for research articles and date lists relevant to 14C and other radioisotopes and techniques used in archaeological, geophysical, oceanographic, and related dating.

www.radiocarbon.org

UNIVERSITY OF BRITISH COLUMBIA

University of British Columbia Library

Primary Unit: Digital Initiatives and Scholarly Communications

Primary Contact:
Allan Bell
Director, Digital Initiatives and Scholarly Communications
604-827-4830
allan.bell@ubc.ca

Website: circle.ubc.ca

PROGRAM OVERVIEW

Mission/description: Digital Initiatives and Scholarly Communication services supports new models of scholarly communications, copyright services, the showcasing of UBC's intellectual output via open access repository services, as well as the digitization of unique historical materials. Digital Initiatives and Scholarly Communication services is a key part of the Library's strategy to support the evolving needs of faculty and students and to support teaching, research and learning at UBC. Our goal is to create sustainable, world-class programs and processes that promote digital scholarship, make UBC research and digital collections openly available to the world, and ensure the long-term preservation of UBC's digital collections.

Year publishing activities began: 2007

Organization: services are distributed across library units/departments

Staff in support of publishing activities (FTE): library staff (4); graduate students (0.3)

Funding sources (%): library operating budget (100)

PUBLISHING ACTIVITIES

Types of publications: faculty-driven journals (7); student-driven journals (8); technical/research reports (47); faculty conference papers and proceedings (80); student conference papers and proceedings (1); newsletters (20); ETDs (1364); undergraduate capstone/honors theses (100); non-thesis graduate student research (55)

Media formats: text; images; audio; video; data

Disciplinary specialties: mining engineering; forestry; education; sustainability; earth and ocean sciences

Top publications: "Guidelines for Mine Haul Road Design" (technical report); "Comparison of Limit States Design" (technical report); "Pain-Enduring Eccentric Exercise" (technical report); "Portable Science: Podcasting as an Outreach Tool for a Large Academic Science and Engineering Library" (technical report); "Wet-Bulb Temperature" (technical report)

Percentage of journals that are peer reviewed: 50

Campus partners: campus departments or programs; individual faculty; graduate students; undergraduate students

Publishing platform(s): DSpace; OJS/OCS/OMP

Digital preservation strategy: Archivematica; COPPUL; LOCKSS; in-house. We participate in the COPPUL LOCKSS PLN.

Additional services: marketing; outreach; training; analytics; metadata; author copyright advisory; digitization; hosting of supplemental content; audio/video streaming

Additional information: We also publish lower-level undergraduate work in our repository, for example, the Science One Program: circle.ubc.ca/handle/2429/6937. Development partner on the Public Knowledge Project (PKP), including the creation and maintenance of user documentation and related training materials, offering hosting and related support, performing testing, participating on PKP's Advisory and Technical committees, and seeking further areas for cooperation.

UNIVERSITY OF CALGARY

Libraries and Cultural Resources

Primary Unit: Centre for Scholarly Communication

Primary Contact:
Tim Au Yeung
Coordinator, Digital Repository Technologies
403-220-8975
ytau@ucalgary.ca

PROGRAM OVERVIEW

Mission/description: The Centre for Scholarly Communication provides innovative solutions for the creation, evaluation, dissemination, and preservation of the research output of the academy. A priority for Libraries and Cultural Resources, the Centre enables scholars through: sustainable electronic publishing using a variety of platforms; robust dissemination of digital collections in multiple formats; a platform for partnerships and discussion of trends and ideas; and solutions for longer term preservation of digital collections.

Year publishing activities began: 2001

Organization: centralized library publishing unit/department

Staff in support of publishing activities (FTE): library staff (2.5)

Funding sources (%): library operating budget (80); grants (20)

PUBLISHING ACTIVITIES

Types of publications: faculty-driven journals (6); student-driven journals (1); journals produced under contract/MOU for external groups (7); technical/research reports (24); faculty conference papers and proceedings (55); student conference papers and proceedings (8); newsletters (3); ETDs (659); occasional papers

Media formats: text; images; audio; video; data

Top publications: *Arctic* (journal); *ARIEL: A Review of International English Literature* (journal); *Journal of Military and Strategic Studies* (journal); ETDs

Percentage of journals that are peer reviewed: 100

Campus partners: campus departments or programs; individual faculty

Other partners: scholarly societies (e.g., Canadian Evaluation Society); research institutes (e.g., Arctic Institute of North America); individual faculty at other Canadian universities (e.g., University of Saskatchewan)

Publishing platform(s): CONTENTdm; DSpace; OJS/OCS/OMP; locally developed software

Digital preservation strategy: Archivematica; COPPUL; DuraCloud/DSpace; Synergies; in-house; digital preservation services under discussion

Additional services: graphic design (print or web); outreach; training; analytics; cataloging; metadata; DOI assignment/allocation of identifiers; author copyright advisory; other author advisory; digitization; hosting of supplemental content; audio/video streaming

Additional information: The university press is a unit within the library. Working collaboratively, the library and press share expertise and technologies to support and extend scholarly publishing services. Changes resulting from the integration include transition of press journals to library-hosted online journals (most now open access) and the initiation of open access book publishing. For this survey, activities associated with books under our Press imprint were not included.

UNIVERSITY OF CALIFORNIA, BERKELEY

Institute for Research on Labor and Employment Library

Primary Unit: The IRLE Library Web Team

Primary Contact:
Terence K. Huwe
Director of Library and Information Resources
510-643-7061
thuwe@library.berkeley.edu

Website: www.irle.berkeley.edu

PROGRAM OVERVIEW

Mission/description: The IRLE Library uses digital technologies to promote the scholarly content created by the Institute for Research on Labor and Employment as well as its affiliated faculty, students, and visiting scholars.

Year publishing activities began: 2002

Organization: individual units create their own library publishing services, but take care to work with the campus-wide and system-wide resources

Staff in support of publishing activities (FTE): library staff (3); graduate students (1); undergraduate students (0.5)

Funding sources (%): library operating budget (60); grants (40)

PUBLISHING ACTIVITIES

Types of publications: faculty-driven journals (1); technical/research reports (60); faculty conference papers and proceedings (20); student conference papers and proceedings (18); newsletters (65); databases (20); In addition to working papers; conference papers and policy reports; GIS web resources

Media formats: text; images; audio; video; data; concept maps/modeling maps/ visualizations; multimedia/interactive content

Disciplinary specialties: employment and wage studies; employment in the "green economy"; public sector labor relations; sociology; management of organizations/organizational behavior

Top publications: "Hidden Cost of Wal-Mart Jobs: Use of Safety Net Programs by Wal-Mart Workers in California" (technical report); "2003 California Establishment Survey: Preliminary Findings on Employer Based Healthcare

Reform" (technical report); "The Impact of San Francisco's Employer Health Spending Requirement: Initial Findings from the Labor and Product Markets" (technical report); "Impact of SB2 on Health Coverage" (technical report)

Percentage of journals that are peer reviewed: 100

Campus partners: campus departments or programs; individual faculty; graduate students

Other partners: California Studies Association

Publishing platform(s): bepress (Digital Commons); DSpace; WordPress; locally developed software

Digital preservation strategy: UC3 Merritt; in-house; digital preservation services under discussion

Additional services: graphic design (print or web); typesetting; copy-editing; marketing; outreach; training; analytics; cataloging; metadata; compiling indexes and/or TOCs; ISSN registration; DOI assignment/allocation of identifiers; dataset management; business model development; budget preparation; contract/license preparation; other author advisory; digitization; hosting of supplemental content

Plans for expansion/future directions: Beginning to use the EPUB format for full-length ebook sales by third party outlets.

UNIVERSITY OF CALIFORNIA SYSTEM
California Digital Library

Primary Unit: Access and Publishing Group

Primary Contact:
Catherine Mitchell
Director, Access and Publishing Group
510-587-6132
catherine.mitchell@ucop.edu

Website: www.escholarship.org

Social media: @eScholarship; facebook.com/eScholarship

PROGRAM OVERVIEW
Mission/description: eScholarship provides a suite of open access, scholarly publishing services and research tools that enable departments, research units, publishing programs, and individual scholars associated with the University of California to have direct control over the creation and dissemination of the full range of their scholarship.

Year publishing activities began: 2002

Organization: centralized library publishing unit/department

Staff in support of publishing activities (FTE): library staff (5)

Funding sources (%): library operating budget (100)

PUBLISHING ACTIVITIES
Types of publications: faculty-driven journals (28); student-driven journals (31); monographs (159); technical/research reports (18607); faculty conference papers and proceedings (967); student conference papers and proceedings (35); ETDs (5354); undergraduate capstone/honors theses (35)

Media formats: text; images; audio; video; data

Disciplinary specialties: law; Romance languages/classics; environmental studies; architecture/urban planning; linguistics/literary studies

Top publications: "Assessing the Future Landscape of Scholarly Communication: An Exploration of Faculty Values and Needs in Seven Disciplines" (technical/ research report); *Dermatology Online Journal* (journal); *Journal of Transnational American Studies* (journal); *Western Journal of Emergency Medicine* (journal); *The Traffic in Praise: Pindar and the Poetics of Social Economy* (monograph)

Percentage of journals that are peer reviewed: 90

Campus partners: UC Press; campus departments or programs; individual faculty; graduate students; undergraduate students

Other partners: Public Knowledge Project; 10 UC Campus Libraries; PubMed; BioMed Central

Publishing platform(s): OJS; locally developed software

Digital preservation strategy: UC3 Merritt

Additional services: outreach; training; analytics; cataloging; DOI assignment/ allocation of identifiers; open URL support; dataset management; peer review management; contract/license preparation; author copyright advisory; other author advisory; digitization; hosting of supplemental content; audio/video streaming

Plans for expansion/future directions: Identify opportunities to support new modes of research by investigating the needs of digital humanities scholars; explore to what extent altmetrics and commenting/annotation provide utility to researchers in different disciplines by experimenting with the provision of related tools and technologies; improve the quality of eScholarship journals by providing baseline standards and guidance regarding best practices for OA publications; empower eScholarship contributors to better understand and manage their copyright and publishing choices; improve the ability of eScholarship research units to more robustly interact with eScholarship by completing an administrative interface project (begun in 2012–13) that provides them with expanded capabilities to control their publication environment within eScholarship; continue to build relationships with and contribute to the broader digital library publishing community via our major development partnership with the Public Knowledge Project; develop and formalize user community engagement processes for Access and Publishing services in order to leverage super-user knowledge/ practices, better align development priorities with user needs, raise awareness of new features/development agenda, work more directly with campus contacts and increase outreach opportunities to new users.

UNIVERSITY OF CENTRAL FLORIDA

John C. Hitt Library

Primary Unit: Information Technology and Digital Initiatives

Primary Contact:
Lee Dotson
Digital Initiatives Librarian
407-823-1236
lee.dotson@ucf.edu

PROGRAM OVERVIEW

Mission/description: The UCF Libraries currently provides publishing support for honors theses, graduate ETDs, and UCF affiliated or UCF faculty-edited open access e-journals. Efforts to support broader dissemination of scholarship include enabling access to a wide audience through freely accessible databases and using Open Journal Systems (OJS) open source publishing software to publish electronic journals from scratch and host electronic journals in Florida OJ. The UCF Libraries collaborates with the Florida Virtual Campus to provide these services.

Year publishing activities began: 2004

Organization: services are distributed across library units/departments

PUBLISHING ACTIVITIES

Types of publications: faculty-driven journals (1); ETDs (260); undergraduate capstone/honors theses (188)

Media formats: text

Campus partners: campus departments or programs; individual faculty

Other partners: Florida Virtual Campus

Publishing platform(s): OJS/OCS/OMP; locally developed software

Digital preservation strategy: FCLA DAITSS

Additional services: outreach; training; analytics; cataloging; metadata; hosting of supplemental content

UNIVERSITY OF COLORADO ANSCHUTZ MEDICAL CAMPUS

Health Sciences Library

Primary Contact:
Heidi Zuniga
Electronic Resources Librarian
303-724-2134
heidi.zuniga@ucdenver.edu

PROGRAM OVERVIEW

Mission/description: The University of Colorado Anschutz Medical Campus Digital Repository will reflect the University's excellence; support the rapid dissemination of research; foster at all levels understanding and appreciation of the value of research, learning, and teaching at CU Anschutz Medical Campus; ensure future, persistent, and reliable access to intellectual assets.

Year publishing activities began: 2012

Organization: services are distributed across several campuses

Staff in support of publishing activities (FTE): library staff (2)

Funding sources (%): library operating budget (100)

PUBLISHING ACTIVITIES

Types of publications: faculty-driven journals (1); student-driven journals (1); technical/research reports (1); faculty conference papers and proceedings (1); ETDs (41)

Media formats: text; images; audio; video; data; multimedia/interactive content

Disciplinary specialties: health sciences

Top publications: ETDs

Percentage of journals that are peer reviewed: 0

Campus partners: individual faculty

Publishing platform(s): DigiTool by ExLibris

Digital preservation strategy: digital preservation services under discussion

Additional services: marketing; outreach; cataloging; metadata; author copyright advisory; digitization

Additional information: We don't consider ourselves to be a "library as publisher" institution at this point, but we certainly do disseminate ETDs and other resources.

Plans for expansion/future directions: Publishing works from recipients of an Open Access Journal Fund program, also administered by our library, which helps authors pay for OA costs; seeing growth in research datasets, and other material that doesn't normally get published but may be of value to researchers; monitoring the publication output of our researchers and trying to direct those articles toward the repository.

UNIVERSITY OF COLORADO DENVER

Auraria Library

Primary Unit: Special Collections and Digital Initiatives

Primary Contact:
Matthew Mariner
Head of Special Collections and Digital Initiatives
303-556-5817
matthew.mariner@ucdenver.edu

Website: digitool.library.colostate.edu/R/?func=collections&collection_id=2379

PROGRAM OVERVIEW

Mission/description: The mission of the Auraria Digital Library Program is to securely host, faithfully present, and freely distribute cultural, historical, educational, and scholarly content to Auraria Campus constituents and the interested public. The curation of scholarly publications, or the intellectual output of Auraria Campus staff, faculty, and students is of particular importance as it serves to promote and legitimize the activities of our institutions amongst our peers.

Year publishing activities began: 2012

Organization: services are distributed across campus

Staff in support of publishing activities (FTE): library staff (2)

Funding sources (%): library operating budget (100)

PUBLISHING ACTIVITIES

Types of publications: ETDs (152)

Media formats: text; images; audio; video

Campus partners: campus departments or programs

Other partners: University of Colorado Denver Graduate School

Publishing platform(s): DigiTool

Digital preservation strategy: Amazon Glacier

Additional services: author copyright advisory; other author advisory; digitization; audio/video streaming

Additional information: Auraria Library actually serves three unaffiliated schools on one campus (CU Denver; Metropolitan State University of Denver; and Community College of Denver). Currently, only CU Denver grants graduate degrees requiring a thesis or dissertation, but said school recently made ETDs mandatory. These are submitted to ProQuest, but co-delivered to the library, where they are hosted and made publicly available. We hope to add more capacity for inclusion of undergraduate works (capstones, undergrad research) that would be published solely in our repository (unlike ETDs, which are technically also held by ProQuest). In addition to these activities, our Scholarly Communications Librarian Jeffrey Beall offers advice to faculty regarding publishing, but he is currently forming plans to offer these services more concretely and publicly (i.e., campus-wide).

Plans for expansion/future directions: Offering a space for unpublished undergraduate works, which are often ignored, but given Auraria's diverse and undergraduate-focused constituency, demand emphasis.

UNIVERSITY OF FLORIDA

George A. Smathers Libraries

Primary Unit: Digital Library Center
ufdc@uflib.ufl.edu

Primary Contact:
Judy Russell
Dean of University Libraries
352-273-2505
jcrussell@ufl.edu

Website: digital.uflib.ufl.edu; ufdc.ufl.edu

PROGRAM OVERVIEW

Organization: services are distributed across library units/departments

PUBLISHING ACTIVITIES

Types of publications: faculty-driven journals (1); student-driven journals (1); journals produced under contract/MOU for external groups (4); newsletters (1); ETDs (1152); databases (14)

Media formats: text; images; audio; video; data; concept maps/modeling maps/ visualizations

Disciplinary specialties: Caribbean studies; entomology; African studies; psychology; physical therapy

Top publications: ARL PD Bank (database); Vodou Archive (digital scholarship database and archive); *African Studies Quarterly* (journal); *Interamerican Journal of Psychology* (journal); *Florida Entomologist* (journal); *Journal of Undergraduate Research* (journal)

Percentage of journals that are peer reviewed: 100

Campus partners: campus departments or programs; individual faculty

Other partners: Florida Virtual Campus (FLVC); Internet Archive; Digital Library of the Caribbean (dLOC); University Press of Florida; Florida Museum of Natural History

Publishing platform(s): OJS/OCS/OMP; locally developed software (SobekCM)

Digital preservation strategy: FCLA DAITSS; in-house

Additional services: outreach; analytics; cataloging; metadata; dataset management; author copyright advisory; digitization; hosting of supplemental content

UNIVERSITY OF GEORGIA

University of Georgia Libraries

Primary Unit: Digital Library of Georgia

Primary Contact:
Andy Carter
Digital Projects Archivist
706-583-0209
cartera@uga.edu

PROGRAM OVERVIEW

Mission/description: Our general objectives are to identify valuable, but overlooked, work from faculty and students, and increase the amount of UGA's scholarly output that is available via open access.

Year publishing activities began: 2010

Organization: centralized library publishing unit/department

Staff in support of publishing activities (FTE): library staff (2)

PUBLISHING ACTIVITIES

Types of publications: faculty-driven journals (1)

Media formats: text; images; audio

Disciplinary specialties: higher education

Campus partners: campus departments or programs; individual faculty

Publishing platform(s): DSpace; OJS/OCS/OMP

Additional services: cataloging; metadata; DOI assignment/allocation of identifiers; author copyright advisory; digitization

Plans for expansion/future directions: Fine-tuning ETD platform; expanding journal hosting efforts using OJS, depending on need and interest on campus.

UNIVERSITY OF GUELPH

University of Guelph Library

Primary Unit: Research Enterprise and Scholarly Communication

Primary Contact:
Wayne Johnston
Head, Research Enterprise and Scholarly Communication
519-824-4120 ext. 56900
wajohnst@uoguelph.ca

PROGRAM OVERVIEW

Mission/description: We seek to disseminate and preserve the scholarly output of the university. We believe open access, both green (self-archiving) and gold (open access journals), is critical to this objective. More broadly, we also seek to promote the digitization and dissemination of Canadian scholarly journal content.

Year publishing activities began: 2004

Organization: centralized library publishing unit/department

Staff in support of publishing activities (FTE): library staff (7); undergraduate students (2)

Funding sources (%): library operating budget (100)

PUBLISHING ACTIVITIES

Types of publications: faculty-driven journals (5); student-driven journals (4); journals produced under contract/MOU for external groups (5); technical/research reports (422); faculty conference papers and proceedings (6); newsletters (2); databases (17); ETDs (1030)

Media formats: text; images; audio; video; data

Disciplinary specialties: agriculture; veterinary sciences; arts; history; international development

Top publications: *Critical Studies in Improvisation* (journal); *International Review of Scottish Studies* (journal); *Partnership: the Canadian Journal of Library and Information Practice and Research* (journal); *Synergies Canada* (journal); *Studies by Undergraduate Researchers at Guelph* (journal)

Percentage of journals that are peer reviewed: 100

Campus partners: campus departments or programs; individual faculty; graduate students; undergraduate students

Other partners: scholarly societies; national organizations; provincial consortia

Publishing platform(s): DSpace; Fedora; OJS/OCS/OMP; DataVerse

Digital preservation strategy: DuraCloud/DSpace; Scholars Portal; Synergies

Additional services: graphic design (print or web); marketing; outreach; training; analytics; cataloging; metadata; dataset management; author copyright advisory; other author advisory; audio/video streaming

UNIVERSITY OF HAWAII AT MANOA

University of Hawaii at Manoa Libraries

Primary Unit: Desktop Network Services

Primary Contact:
Beth Tillinghast
Web Support Librarian, Institutional Repositories Manager
808-956-6130
betht@hawaii.edu

PROGRAM OVERVIEW

Mission/description: Though the University of Hawaii at Manoa currently does not have a formal library publishing program, our library is involved in providing publishing services through the various collections hosted in our institutional repository, ScholarSpace. We provide the hosting services for numerous department journal publications, conference proceedings, technical reports, department newsletters, as well as open access to some dissertations and theses. The publishing activities are consistent with our mission of acquiring, organizing, preserving, and providing access to information resources vital to the learning, teaching, and research mission of the University of Hawaii at Manoa.

Year publishing activities began: 2007

Organization: services are distributed across campus

Staff in support of publishing activities (FTE): library staff (0.1); graduate students (0.2)

Funding sources (%): library operating budget (90); non-library campus budget (5); charge backs (5)

PUBLISHING ACTIVITIES

Types of publications: faculty-driven journals (5); student-driven journals (1); technical/research reports (20); faculty conference papers and proceedings (5); newsletters (12); databases (5); ETDs (3000); undergraduate capstone/honors theses (1); datasets

Media formats: text; images; audio; video; data

Disciplinary specialties: language documentation; social work; entomology; Pacific Islands culture; Southeast Asian culture

Top publications: *Language Documentation and Conservation* (journal); *Ethnobotany Research and Applications* (journal); *The Contemporary Pacific* (journal); *Journal of Indigenous Social Development* (journal); *Explorations* (journal)

Percentage of journals that are peer reviewed: 100

Campus partners: campus departments or programs; individual faculty; graduate students

Publishing platform(s): DSpace

Digital preservation strategy: Archive-It; Portico; in-house

Additional services: DOI assignment/allocation of identifiers; dataset management; author copyright advisory; digitization; hosting of supplemental content

UNIVERSITY OF IDAHO

University of Idaho Library

Primary Unit: Digital Initiatives

Primary Contact:
Devin Becker
Digital Initiatives Librarian
208-885-7040
dbecker@uidaho.edu

Website: www.lib.uidaho.edu/digital; journals.lib.uidaho.edu

PROGRAM OVERVIEW

Mission/description: The Digital Initiatives department works to preserve and make accessible publications and other research products from researchers and affiliates of the University of Idaho via its open access publishing capabilities.

Year publishing activities began: 2013

Organization: centralized library publishing unit/department

Staff in support of publishing activities (FTE): library staff (1.5); undergraduate students (0.5)

Funding sources (%): library operating budget (40); grants (60)

PUBLISHING ACTIVITIES

Types of publications: student-driven journals (1); databases (4)

Media formats: text; images; audio; video; concept maps/modeling maps/ visualizations; multimedia/interactive content

Disciplinary specialties: rangeland ecology and management; creative writing

Top publications: *Fugue* (journal); *Journal of Rangeland Applications* (journal)

Percentage of journals that are peer reviewed: 100

Campus partners: campus departments or programs

Publishing platform(s): CONTENTdm; OJS/OCS/OMP

Digital preservation strategy: in-house

Additional services: graphic design (print or web); typesetting; outreach; training; analytics; cataloging; metadata; compiling indexes and/or TOCs; DOI assignment/allocation of identifiers; open URL support; digitization; hosting of supplemental content

Additional information: We also publish a number of digital collections of historical images and documents.

Plans for expansion/future directions: Bringing ETDs online; using ETDs to start developing a more robust (and visible) institutional repository.

UNIVERSITY OF ILLINOIS AT CHICAGO

University Library

Primary Unit: Scholarly Communications
escholarship@uic.edu

Primary Contact:
Sandy De Groote
Scholarly Communications Librarian
312-413-9494
sgroote@uic.edu

Website: library.uic.edu/home/services/escholarship

PROGRAM OVERVIEW

Mission/description: The objective/mission of the UIC University Library publishing program is to advance scholarly knowledge in a cost-effective manner.

Year publishing activities began: 2007

Organization: centralized library publishing unit/department

Staff in support of publishing activities (FTE): library staff (1); undergraduate students (0.5)

Funding sources (%): library operating budget (95); charge backs (5)

PUBLISHING ACTIVITIES

Types of publications: faculty-driven journals (5); student-driven journals (1); newsletters (1); ETDs (700)

Media formats: text; images; data

Disciplinary specialties: social work; Internet studies; public health informatics

Top publications: *First Monday* (journal); *Online Journal of Public Health Informatics* (journal); *Behavior and Social Issues* (journal); *Uncommon Culture* (journal); *Journal of Biomedical Discovery and Collaboration* (journal)

Percentage of journals that are peer reviewed: 80

Campus partners: individual faculty

Publishing platform(s): CONTENTdm; DSpace; OJS/OCS/OMP; Inera eXtyles

Digital preservation strategy: HathiTrust; LOCKSS

Additional services: graphic design (print or web); typesetting; marketing; training; metadata; notification of A&I sources; ISSN registration; DOI assignment/allocation of identifiers; dataset management; author copyright advisory; digitization; Word to XML conversion

Plans for expansion/future directions: Exploring monograph publishing.

HIGHLIGHTED PUBLICATION

First Monday

This month: September 2013

Also this month!

Announcements

First Monday is one of the first openly accessible, peer–reviewed journals on the Internet, solely devoted to the Internet.

firstmonday.org/index

UNIVERSITY OF IOWA

University of Iowa Libraries

Primary Unit: Digital Research and Publishing
lib-ir@uiowa.edu

Primary Contact:
Wendy Robertson
Digital Scholarship Librarian
319-335-5821
wendy-robertson@uiowa.edu

Website: www.lib.uiowa.edu/drp/publishing

Social media: @IowaResO

PROGRAM OVERVIEW

Mission/description: Digital Research and Publishing explores ways that academic libraries can best leverage digital collections, resources, and expertise to support faculty and student scholars by: collaborating on interdisciplinary scholarship built upon digital collections; offering publishing services to support sustainable scholarly communication; engaging the community through participatory digital initiatives; promoting widespread use and reuse of locally built repositories and archives; and advancing new technologies that support digital research and publishing.

Year publishing activities began: 2009

Organization: services are distributed across library units/departments

Staff in support of publishing activities (FTE): library staff (1.5); graduate students (0.5)

Funding sources (%): library operating budget (100)

PUBLISHING ACTIVITIES

Types of publications: faculty-driven journals (6); student-driven journals (2); journals produced under contract/MOU for external groups (1); technical/ research reports (9); faculty conference papers and proceedings (1); newsletters (1); ETDs (530)

Media formats: text

Top publications: *Walt Whitman Quarterly Review* (journal); *Medieval Feminist Forum* (journal); *Proceedings in Obstetrics & Gynecology* (journal); *Iowa Journal of Cultural Studies* (journal); *Poroi* (journal)

Percentage of journals that are peer reviewed: 90

Campus partners: campus departments or programs; individual faculty; graduate students

Other partners: Society for Medieval Feminist Scholarship

Publishing platform(s): bepress (Digital Commons); CONTENTdm; WordPress

Digital preservation strategy: Archive-It; LOCKSS; in-house; digital preservation services under discussion

Additional services: cataloging; metadata; notification of A&I sources; ISSN registration; DOI assignment/allocation of identifiers; peer review management; digitization; hosting of supplemental content; audio/video streaming

Plans for expansion/future directions: Working on adding additional services, such as DOIs, HTML versions of articles, and possibly some formatting of content; assessing campus needs for datasets.

UNIVERSITY OF KANSAS

KU Libraries

Library Publishing Coalition — Contributing Institution

Primary Unit: Center for Faculty Initiatives and Engagement
kuscholarworks@ku.edu

Primary Contact:
Marianne Reed
Digital Information Specialist
785-864-8913
mreed@ku.edu

Website: journals.ku.edu

PROGRAM OVERVIEW

Mission/description: Digital Publishing Services provides support to the KU community for the design, management, and distribution of online publications, including journals, conference proceedings, monographs, and other scholarly content. We help scholars explore new and emerging publishing models in our changing scholarly communication environment, and we help monitor and address campus concerns and questions about electronic publishing. These services are intended to enable online publishing for campus publications, and help make their content available in a manner that promotes increased visibility and access, and ensures long-term stewardship of the materials.

Year publishing activities began: 2007

Organization: centralized library publishing unit/department; transitioning to distribution across library units

Staff in support of publishing activities (FTE): library staff (0.25); undergraduate students (0.05)

Funding sources (%): library operating budget (100)

PUBLISHING ACTIVITIES

Types of publications: faculty-driven journals (15); student-driven journals (1); monographs (7); faculty conference papers and proceedings (2); ETDs (214); undergraduate capstone/honors theses (70); occasional lectures; oral histories and interviews

Media formats: text; audio; video

Disciplinary specialties: philosophy; natural science; humanities; oral history and interviews; linguistics

Top publications: *Biodiversity Informatics* (journal); *American Studies* (journal); *Latin American Theater Review* (journal); Kansas Working Papers in Linguistics (working papers); Treatise Online (preprints)

Percentage of journals that are peer reviewed: 100

Campus partners: individual faculty; graduate students

Publishing platform(s): DSpace; OJS/OCS/OMP; XTF

Digital preservation strategy: Portico; digital preservation services under discussion

Additional services: outreach; training; analytics; cataloging; metadata; author copyright advisory; digitization; hosting of supplemental content; audio/video streaming; ISBNs; consulting on publishing models and issues

Plans for expansion/future directions: Some services are ongoing. A strategic initiative to expand the program is pending.

UNIVERSITY OF KENTUCKY

University of Kentucky Libraries

Primary Unit: Department of Digital Scholarship
UKnowledge@lsv.uky.edu

Primary Contact:
Adrian K. Ho
Director of Digital Scholarship
859-218-0895
adrian.ho@uky.edu

Website: uknowledge.uky.edu

PROGRAM OVERVIEW

Mission/description: The University of Kentucky (UK) Libraries launched an institutional repository (UKnowledge) in late 2010 to champion the integration and transformation of scholarly communication within the UK community. The initiative sought to improve access by students, faculty, and researchers to appropriate resources for maximizing the dissemination of their research and scholarship in an open and digital environment. A crucial component of UKnowledge is providing publishing services to broadly disseminate scholarship created or sponsored by the UK community. We provide a flexible platform to publish a variety of scholarly content and to expand the discoverability of the published works. Additionally, we are establishing a separate digital repository for the long-term preservation of the published content and research datasets. Using state-of-the-art technologies, we are able to offer campus constituents sought-after services in different stages of the scholarly communication life cycle to help them thrive and succeed. We also inform them of scholarly communication issues such as open access, author rights, and the economics of journal publishing. Providing library publishing services is one avenue through which we are making significant contributions to the fulfillment of UK's mission.

Year publishing activities began: 2010

Organization: services are distributed across library units/departments

Staff in support of publishing activities (FTE): library staff (2)

Funding sources (%): library materials budget (99); charge backs (1)

PUBLISHING ACTIVITIES

Types of publications: faculty-driven journals (2); student-driven journals (2); technical/research reports (9); faculty conference papers and proceedings (4); newsletters (1); ETDs (2040); undergraduate capstone/honors theses (6); image galleries/virtual exhibits (10)

Media formats: text; images

Disciplinary specialties: higher education; Hispanic studies; public health; undergraduate research (multidisciplinary)

Top publications: *Kentucky Journal of Higher Education Policy and Practice* (journal); *Nomenclatura: Aproximaciones a Los Estudios Hispánicos* (journal); *Frontiers in Public Health Services and Systems Research* (journal); *Kaleidoscope: The University of Kentucky Journal of Undergraduate Scholarship* (journal)

Percentage of journals that are peer reviewed: 75

Campus partners: campus departments or programs; individual faculty; graduate students

Publishing platform(s): bepress (Digital Commons)

Digital preservation strategy: in-house

Additional services: graphic design (print or web); training; analytics; cataloging; metadata; notification of A&I sources; ISSN registration; open URL support; peer review management; contract/license preparation; author copyright advisory; other author advisory; digitization; hosting of supplemental content

HIGHLIGHTED PUBLICATION

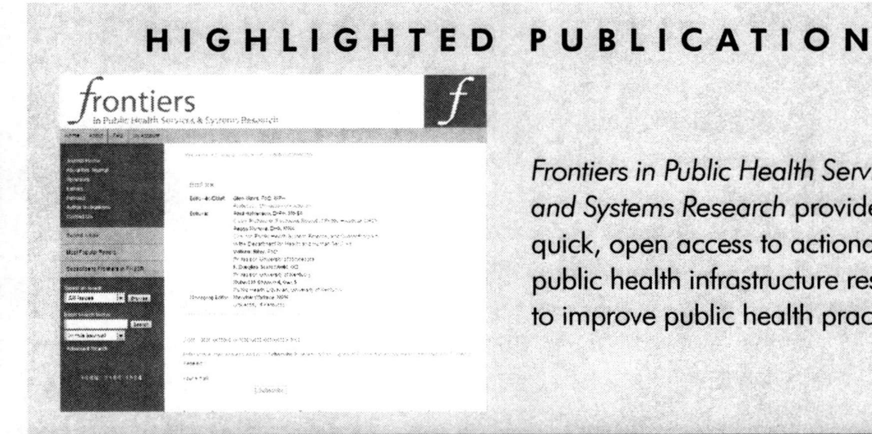

Frontiers in Public Health Services and Systems Research provides quick, open access to actionable public health infrastructure research to improve public health practices.

uknowledge.uky.edu/frontiersinphssr

Plans for expansion/future directions: Strengthening existing library publishing partnerships; bringing more campus constituents on board; building upon our current library publishing services (e.g., partnering with the UK Graduate School to complete the integration of our library publishing services into the workflow as they implement an electronic thesis and dissertation mandate); pursuing additional opportunities to collaborate with various campus units in support of undergraduate research as we celebrate UK students' academic achievements by making them visible and accessible worldwide; assisting UK-based print journals to create their online presence and extend their reach beyond academia; exploring data publishing in partnership with UK researchers; continuing to advocate open access and open licensing as well as inform the UK community of new scholarly communication practices such as alternative metrics, open peer review, and researcher identity management; making UKnowledge the primary online publishing avenue for UK-based research and scholarship.

UNIVERSITY OF MARYLAND COLLEGE PARK
McKeldin Library

Primary Unit: Digital Stewardship

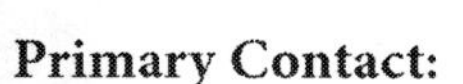

Primary Contact:
Terry M. Owen
DRUM Coordinator
301-314-1328
towen@umd.edu

Website: publish.lib.umd.edu; drum.lib.umd.edu

PROGRAM OVERVIEW

Mission/description: Capture, preserve, and provide access to the output of University of Maryland faculty, researchers, centers, and labs.

Year publishing activities began: 2004

Organization: centralized library publishing unit/department

Staff in support of publishing activities (FTE): library staff (1); graduate students (0.5)

Funding sources (%): library operating budget (100)

PUBLISHING ACTIVITIES

Types of publications: faculty-driven journals (1); student-driven journals (1); technical/research reports (124); newsletters (1); ETDs (946); undergraduate capstone/honors theses (24)

Campus partners: campus departments or programs; individual faculty; graduate students

Publishing platform(s): DSpace; OJS/OCS/OMP

Digital preservation strategy: in-house; digital preservation services under discussion

Additional services: analytics; metadata; ISSN registration; author copyright advisory; hosting of supplemental content

Plans for expansion/future directions: Expanding into ePublishing in 2013, including faculty and student-produced e-publications.

UNIVERSITY OF MASSACHUSETTS AMHERST

W.E.B. Du Bois Library

Library Publishing Coalition
FOUNDING INSTITUTION

Primary Unit: Office of Scholarly Communication
scholarworks@library.umass.edu

Primary Contact:
Marilyn S. Billings
Scholarly Communication & Special Initiatives Librarian
413-545-6891
mbillings@library.umass.edu

Website: scholarworks.umass.edu

PROGRAM OVERVIEW

Mission/description: ScholarWorks@UMass Amherst, an open access digital repository service, was established in 2006 to provide a digital showcase of the unique research and scholarly outputs of members of the University of Massachusetts Amherst community. It provides a platform for the distribution of content such as electronic dissertations, master's theses, and capstone projects as well as scholarly output of academic departments, research centers, and institutes. ScholarWorks provides a wide variety of scholarly publishing services including: online journal publishing and conference management system; collaboration with scholarly presses to provide permanent location and URLs for supplementary content for scholarly monographs, texts, and other scholarly materials. ScholarWorks provides many services for research support that can be used in conjunction with grant applications, which now require applicants to detail how the results of the funded research will be showcased and disseminated. The ScholarWorks service can be included as part of the overall data management strategy for research results, reports, new journal services, conference proceedings, etc. These value-added services enhance the professional visibility for faculty and researchers and provide excellent search and retrieval facilities and broader dissemination as well as increased use of materials through services such as Google Scholar and other Internet search engines.

Year publishing activities began: 2006

Organization: centralized library publishing unit/department

Staff in support of publishing activities (FTE): library staff (2.5); undergraduate students (0.5)

Funding sources (%): library materials budget (20); library operating budget (70); non-library campus budget (10)

PUBLISHING ACTIVITIES

Types of publications: faculty-driven journals (7); student-driven journals (1); journals produced under contract/MOU for external groups (5); monographs (8); textbooks; (2); technical/research reports (1000); faculty conference papers and proceedings (100); newsletters (2); ETDs (3200); graduate student capstones and practicums

Media formats: text; images; audio; video; data

Disciplinary specialties: anthropology; engineering; community engagement; nursing; hospitality and tourism

Top publications: "How To Do Case Study Research" (technical report); "The Impact of Language Barrier & Cultural Differences on Restaurant Experiences: A Grounded Theory Approach" (conference proceedings); "Theme Park Development Costs: Initial Investment Cost Per First Year Attendee" (conference proceedings); "The Form of the Preludes to Bach's Unaccompanied Cello Suites" (thesis); "Ratio Analysis for the Hospitality Industry: A Cross Sector Comparison of Financial Trends in the Lodging, Restaurant, Airline and Amusement Sectors" (journal article)

Percentage of journals that are peer reviewed: 100

Campus partners: campus departments or programs; individual faculty; graduate students

Publishing platform(s): bepress (Digital Commons); EPrints; Fedora

Digital preservation strategy: LOCKSS; in-house; digital preservation services under discussion

Additional services: graphic design (print or web); marketing; outreach; training; analytics; cataloging; metadata; ISSN registration; DOI assignment/allocation of identifiers; open URL support; dataset management; peer review management; contract/license preparation; author copyright advisory; other author advisory; digitization; hosting of supplemental content

Additional information: We are looking into the possibilities of coordinating more closely with our University Press on a variety of services. They have expertise but not the additional time to assist with the types of publishing services faculty are starting to ask for, such as copy-editing, proofing. We are also members of the Networked Digital Library of Theses and Dissertations (NDLTD).

Plans for expansion/future directions: Exploring additional publication services in collaboration with other groups on campus (copy-editing, proofing, graphic design, referral services); engaging in more extensive collaboration with the Office of Research on data management, intellectual property/copyright; and expanding into capturing undergraduate student work/projects.

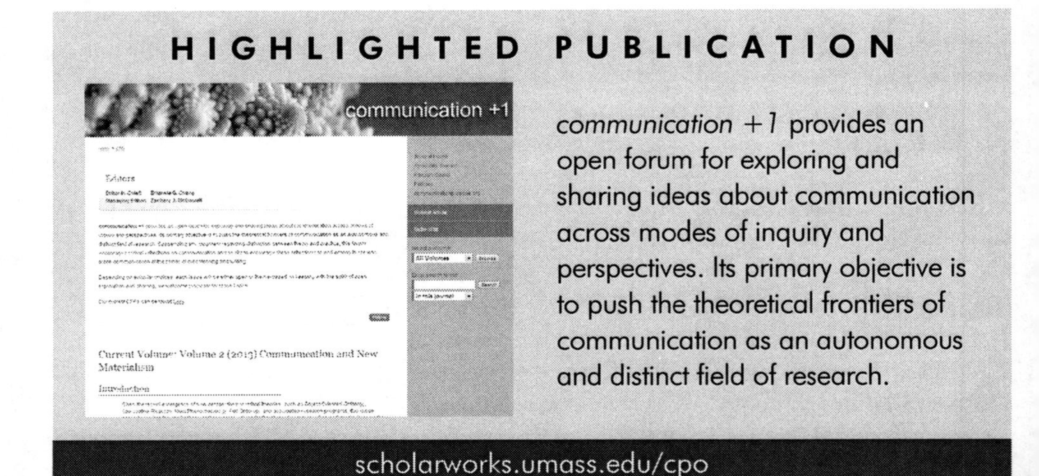

UNIVERSITY OF MASSACHUSETTS MEDICAL SCHOOL

Lamar Soutter Library

Library Publishing Coalition

Primary Unit: Research and Scholarly Communication Services

Primary Contact:
Rebecca Reznik-Zellen
Head of Research & Scholarly Communication Services
508-856-6810
Rebecca.Reznik-Zellen@umassmed.edu

Website: escholarship.umassmed.edu/about.html

PROGRAM OVERVIEW

Mission/description: eScholarship@UMMS is a digital repository offering worldwide access to the research and scholarly work of the University of Massachusetts Medical School community. The goal is to bring together the University's scholarly output in order to enhance its visibility and accessibility. We help individual researchers and departments organize and publicize their research beyond the walls of the Medical School, archiving publications, posters, presentations, and other materials they produce in their scholarly pursuits. Our publishing services—including the *Journal of eScience Librarianship* and two other open access peer-reviewed electronic journals, student dissertations and theses, and conference proceedings—highlight the works of University of Massachusetts Medical School authors and others.

Year publishing activities began: 2007

Organization: services are distributed across library units/departments

Staff in support of publishing activities (FTE): library staff (1)

Funding sources (%): library operating budget (75); grants (25)

PUBLISHING ACTIVITIES

Types of publications: faculty-driven journals (3); monographs (1); textbooks; (1); faculty conference papers and proceedings (125); student conference papers and proceedings (49); ETDs (77); finding aids (7); book reviews (8)

Media formats: text; images; audio; video; multimedia/interactive content

Disciplinary specialties: library science; psychiatry/mental health research; neurology; clinical and translational science; life sciences

Top publications: *Journal of eScience Librarianship* (journal); ETDs; *Psychiatry Information in Brief* (journal); *Neurological Bulletin* (journal); *A History of the University of Massachusetts Medical School* (e-book)

Percentage of journals that are peer reviewed: 100

Campus partners: campus departments or programs; individual faculty; graduate students

Publishing platform(s): bepress (Digital Commons)

Digital preservation strategy: in-house; digital preservation services under discussion

Additional services: copy-editing; marketing; outreach; training; metadata; ISSN registration; DOI assignment/allocation of identifiers; open URL support; dataset management; peer review management; author copyright advisory; hosting of supplemental content; audio/video streaming; altmetrics data

Plans for expansion/future directions: Expanding our publishing services to additional departments within the medical school, incorporating more multimedia, and enhancing publications with altmetrics data.

UNIVERSITY OF MICHIGAN

University Library

Primary Unit: Michigan Publishing
mpublishing@umich.edu

Website: www.publishing.umich.edu

Social media: @M_Publishing

PROGRAM OVERVIEW

Mission/description: Michigan Publishing is the hub of scholarly publishing at the University of Michigan, and is a part of its dynamic and innovative University Library. Our mission as publishers, librarians, copyright experts, and technologists is to support the communications needs of scholars, and to publish, promote, and preserve the scholarly record.

Year publishing activities began: 2001

Organization: centralized library publishing unit/department

Funding sources (%): library operating budget (60); sales revenue (40)

PUBLISHING ACTIVITIES

Campus partners: campus departments or programs; individual faculty; graduate students; undergraduate students

Other partners: Open Humanities Press; American Council of Learned Societies

Publishing platform(s): DSpace; WordPress; locally developed software

Digital preservation strategy: HathiTrust; in-house

Additional services: typesetting; copy-editing; marketing; outreach; training; analytics; cataloging; metadata; ISSN registration; DOI assignment/allocation of identifiers; contract/license preparation; author copyright advisory; other author advisory

UNIVERSITY OF MINNESOTA

University of Minnesota Libraries

Primary Unit: Content and Collections Division
jkirchne@umn.edu

Primary Contact:
Joy Kirchner
AUL for Content & Collections
612-624-2312
jkirchne@umn.edu

PROGRAM OVERVIEW

Year publishing activities began: 2006

Organization: services are distributed across library units/departments

Funding sources (%): library operating budget (50); endowment income (25); grants (25)

PUBLISHING ACTIVITIES

Types of publications: faculty conference papers and proceedings (2520); ETDs (680); working papers; blogs; online dictionary

Media formats: text; images; audio; video; data

Campus partners: individual faculty

Publishing platform(s): CONTENTdm; DSpace; MovableType; Drupal

Digital preservation strategy: CLOCKSS; DuraCloud/DSpace; HathiTrust; Portico; Omeka

Additional services: training; analytics; metadata; open URL support; dataset management; contract/license preparation; author copyright advisory; other author advisory; digitization; hosting of supplemental content; audio/video streaming

Additional information: Currently developing a program in a new division.

UNIVERSITY OF NEBRASKA-LINCOLN

University of Nebraska-Lincoln Libraries

Primary Unit: Zea Books/Office of Scholarly Communications
proyster@unl.edu

Primary Contact:
Paul Royster
Publisher, Zea Books
402-472-3628
proyster@unl.edu

Website: digitalcommons.unl.edu/zea

PROGRAM OVERVIEW

Mission/description: Zea Books is the digital and on-demand publishing operation of the University of Nebraska-Lincoln Libraries.

Year publishing activities began: 2005

Organization: centralized library publishing unit/department

Staff in support of publishing activities (FTE): library staff (2)

Funding sources (%): library operating budget (100)

PUBLISHING ACTIVITIES

Types of publications: faculty-driven journals (3); student-driven journals (1); journals produced under contract/MOU for external groups (2); monographs (7); textbooks (1); technical/research reports (1); faculty conference papers and proceedings (4); student conference papers and proceedings (3); newsletters (1); databases (1); ETDs (200); undergraduate capstone/honors theses (12)

Media formats: text; images; data; concept maps/modeling maps/visualizations; multimedia/interactive content

Campus partners: individual faculty

Other partners: Nebraska Academy of Sciences; Center for Great Plains Studies; Textile Society of America; Lester A. Larsen Tractor and Power Museum; Center for Systemic Entomology; Nebraska Ornithological Union

Publishing platform(s): bepress (Digital Commons)

Additional services: graphic design (print or web); typesetting; copy-editing; marketing; outreach; training; analytics; cataloging; metadata; compiling indexes and/or TOCs; notification of A&I sources; ISSN registration; open URL support; peer review management; business model development; contract/license preparation; author copyright advisory; other author advisory; digitization; hosting of supplemental content

UNIVERSITY OF NORTH CAROLINA AT CHAPEL HILL

University Library

Primary Unit: Library Administration

Primary Contact:
Will Owen
Associate University Librarian for Technical Services and Systems
919-962-1301
owen@email.unc.edu

PROGRAM OVERVIEW

Mission/description: The Library has historically published, in print, specialized monographs on topics related to the University or Library. We publish ETDs electronically and provide digital editions and original scholarly interpretations in support of research and instruction with a special emphasis on the American South.

Year publishing activities began: 1995

Organization: services are distributed across library units/departments

Staff in support of publishing activities (FTE): library staff (2); graduate students (0.5)

Funding sources (%): library operating budget (100)

PUBLISHING ACTIVITIES

Types of publications: ETDs (570); digital humanities research projects

Media formats: text; images; audio; video; data; concept maps/modeling maps/visualizations

Disciplinary specialties: the American South

Top publications: Documenting the American South (digital collection)

Campus partners: UNC Press; campus departments or programs; individual faculty; graduate students; undergraduate students

Publishing platform(s): CONTENTdm; Fedora; locally developed software

Digital preservation strategy: Archive-It; HathiTrust; in-house (Carolina Digital Repository); Internet Archive

Additional services: training; cataloging; metadata; author copyright advisory; digitization; hosting of supplemental content

Plans for expansion/future directions: Collaborating with researchers on archiving, preserving, and publishing research data; collaborating with UNC Press for print-on-demand publications.

UNIVERSITY OF NORTH CAROLINA AT CHARLOTTE

Atkins Library

Library Publishing Coalition — Contributing Institution

Primary Unit: Digital Scholarship Lab
atkins-dsl@uncc.edu

Primary Contact:
Somaly Kim Wu
Digital Scholarship Librarian
704-687-1112
skimwu@uncc.edu

Website: journals.uncc.edu; dsl.uncc.edu/dsl/services/publication

PROGRAM OVERVIEW

Mission/description: We support the publication of scholarly journals online and assist journal editors with the management, editorial work, and production of their scholarly journal. The DSL offers journal hosting support services to UNC Charlotte faculty. Our services are built on the Open Journal System (OJS) journal management software that facilitates the publication of online peer-reviewed journals. DSL services include platform software hosting, updates, and copyright consulting.

Year publishing activities began: 2012

Organization: centralized library publishing unit/department

Staff in support of publishing activities (FTE): library staff (1)

Funding sources (%): library operating budget (100)

PUBLISHING ACTIVITIES

Types of publications: faculty-driven journals (3)

Media formats: text

Disciplinary specialties: education; psychology; urban education

Top publications: *NHSA Dialog* (journal); *Urban Education Research and Policy Annuals* (journal); *Undergraduate Journal of Psychology* (journal)

Percentage of journals that are peer reviewed: 100

Campus partners: individual faculty

Publishing platform(s): OJS/OCS/OMP

Digital preservation strategy: digital preservation services under discussion

Additional services: graphic design (print or web); training; ISSN registration; dataset management; author copyright advisory

Plans for expansion/future directions: Building an institutional repository that is planned to be online within the year.

UNIVERSITY OF NORTH CAROLINA AT GREENSBORO

University Libraries

Primary Unit: Collections and Scholarly Communications

Primary Contact:
Beth Bernhardt
Assistant Dean for Collection Management and Scholarly Communications
336-256-1210
brbernha@uncg.edu

PROGRAM OVERVIEW

Mission/description: still in development

Year publishing activities began: 2004

Organization: services are distributed across library units/departments

Staff in support of publishing activities (FTE): library staff (0.5)

Funding sources (%): other (100)

PUBLISHING ACTIVITIES

Types of publications: faculty-driven journals (7); journals produced under contract/MOU for external groups (1); technical/research reports (23); faculty conference papers and proceedings (32); databases (4); ETDs (1978)

Media formats: text; images; audio; video; data; concept maps/modeling maps/ visualizations; multimedia/interactive content

Disciplinary specialties: public health; education; nursing; sociology

Top publications: *International Journal of Nurse Practitioner Educators* (journal); *The International Journal of Critical Pedagogy* (journal); *Journal of Backcountry Studies* (journal); *Journal of Learning Spaces* (journal); *Partnerships: A Journal of Service-Learning and Civic Engagement* (journal)

Percentage of journals that are peer reviewed: 85

Campus partners: campus departments or programs; individual faculty; graduate students; undergraduate students

Publishing platform(s): CONTENTdm; OJS/OCS/OMP; locally developed software

Digital preservation strategy: HathiTrust; in-house; digital preservation services under discussion

Additional services: training; analytics; cataloging; metadata; author copyright advisory; other author advisory; digitization; hosting of supplemental content

Plans for expansion/future directions: Hosting OJS for other regional libraries; supporting faculty in new scholarly media, such as database and UI design, web pages, and usability.

HIGHLIGHTED PUBLICATION

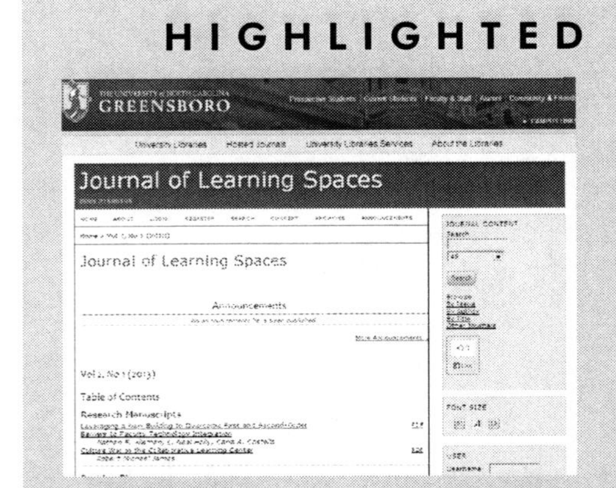

A peer-reviewed, open-access journal published biannually, *The Journal of Learning Spaces* provides a scholarly, multidisciplinary forum for research articles, case studies, book reviews, and position pieces related to all aspects of learning space design, operation, pedagogy, and assessment in higher education.

partnershipsjournal.org/index.php/jls

UNIVERSITY OF NORTH TEXAS

University of North Texas Libraries

Library Publishing Coalition

Primary Unit: Scholarly Publishing Services

Primary Contact:
Martin Halbert
Dean of Libraries
940-565-3025
martin.halbert@unt.edu

PROGRAM OVERVIEW

Mission/description: The UNT Libraries Scholarly Publishing Services are a collaborative program between faculty and the library to develop new and innovative forms of scholarly publications, especially using digital technologies.

Year publishing activities began: 2009

Organization: services are distributed across library units/departments

Staff in support of publishing activities (FTE): library staff (4), graduate students (1)

Funding sources (%): library operating budget (100)

PUBLISHING ACTIVITIES

Types of publications: faculty-driven journals (1); student-driven journals (1)

Media formats: text; images; audio; video; data; multimedia/interactive content

Disciplinary specialties: electronic arts

Top publications: *Möbius Journal* (journal); *The Eagle Feather* (journal)

Percentage of journals that are peer reviewed: 50

Campus partners: campus departments or programs; individual faculty; graduate students; undergraduate students

Other partners: Texas State Historical Association

Publishing platform(s): locally developed software

Digital preservation strategy: digital preservation services under discussion

Additional services: graphic design (print or web); metadata

Plans for expansion/future directions: Cultivate new ideas for collaborative scholarly publications.

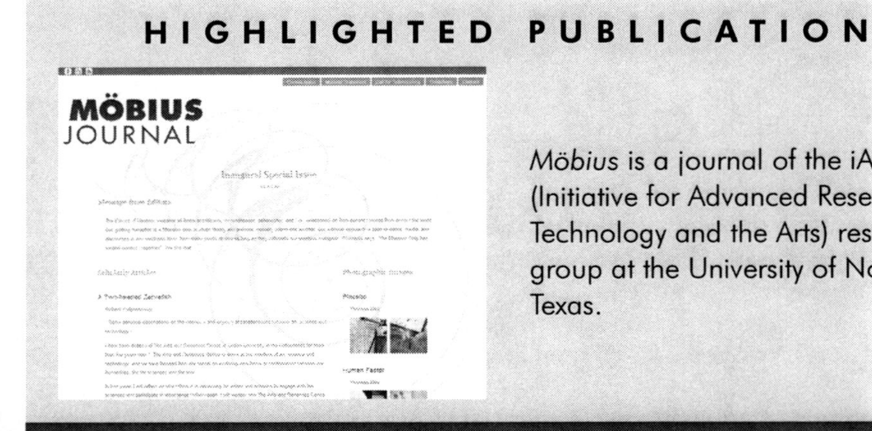

UNIVERSITY OF OREGON

University of Oregon Libraries

Primary Unit: Digital Scholarship Center

Primary Contact:
John Russell
Scholarly Communications Librarian
541-346-2689
johnruss@uoregon.edu

Website: library.uoregon.edu/digitalscholarship

PROGRAM OVERVIEW

Mission/description: The Digital Scholarship Center (DSC) collaborates with faculty and students to transform research and scholarly communication using new media and digital technologies. Based on a foundation of access, sharing, and preservation, the DSC provides digital asset management, digital preservation, training, consultations, and tools for digital scholarship.

Year publishing activities began: 2003

Organization: services are distributed across library units/departments

Staff in support of publishing activities (FTE): library staff (1.25); graduate students (0.4); undergraduate students (0.2)

Funding sources (%): library operating budget (100)

PUBLISHING ACTIVITIES

Types of publications: faculty-driven journals (3); student-driven journals (1); monographs (4); technical/research reports (240); ETDs (230); undergraduate capstone/honors theses (25)

Media formats: text; images; audio; video; data; concept maps/modeling maps/visualizations; multimedia/interactive content

Disciplinary specialties: humanities; gender studies

Top publications: *Ada: A Journal of Gender, New Media, and Technology* (journal); *Konturen* (journal); *Oregon Undergraduate Research Journal* (journal); *Humanist Studies & the Digital Age* (journal)

Percentage of journals that are peer reviewed: 100

Campus partners: campus departments or programs; individual faculty; graduate students; undergraduate students

Other partners: Fembot Collective

Publishing platform(s): CONTENTdm; DSpace; OJS/OCS/OMP; WordPress

Digital preservation strategy: in-house

Additional services: graphic design (print or web); copy-editing; training; analytics; cataloging; metadata; ISSN registration; DOI assignment/allocation of identifiers; open URL support; dataset management; peer review management; contract/license preparation; author copyright advisory; other author advisory; digitization; hosting of supplemental content; audio/video streaming

Plans for expansion/future directions: Increasing quality control over publications.

UNIVERSITY OF PITTSBURGH

University Library System

Library Publishing Coalition — Founding Institution

Primary Unit: Office of Scholarly Communication and Publishing
oscp@mail.pitt.edu

Primary Contact:
Timothy S. Deliyannides
Director, Office of Scholarly Communication and Publishing
412-648-3254
tsd@pitt.edu

Website: www.library.pitt.edu/dscribe

Social media: @OSCP_Pitt

PROGRAM OVERVIEW

Mission/description: The University Library System, University of Pittsburgh offers a full range of publishing services for a variety of content types, specializing in scholarly journals and subject-based open access repositories. Because we are committed to helping research communities share knowledge and ideas through open and responsible collaboration, we subsidize the costs of electronic publishing and provide incentives to promote open access to scholarly research. Our program promotes open access journal publishing at a very low cost; eliminates the high cost of print journal publication and distribution; allows easy collaboration among authors, editors, and reviewers regardless of location; enhances the visibility, searchability, and navigation of publications; and incorporates innovative and sustainable technologies to speed and facilitate scholarly publishing. We are seeking partners around the world who share our commitment to open access to scholarly research information.

Year publishing activities began: 1999

Organization: centralized library publishing unit/department

Staff in support of publishing activities (FTE): library staff (4.5); graduate students (0.5)

Funding sources (%): library operating budget (75); non-library campus budget (5); charge backs (20)

PUBLISHING ACTIVITIES

Types of publications: faculty-driven journals (10); student-driven journals (10); journals produced under contract/MOU for external groups (14); monographs (209); technical/research reports (4587); faculty conference papers and proceedings (273); ETDs (565); undergraduate capstone/honors theses (25); government documents (2340); unpublished article manuscripts (308)

Media formats: text; images; audio; video; data; concept maps/modeling maps/visualizations; multimedia/interactive content

Disciplinary specialties: Latin American studies; European studies; history and philosophy of science; law; health sciences

Top publications: *Revista Iberoamericana* (journal); *University of Pittsburgh Law Review* (journal); *International Journal of Telerehabilitation* (journal); Archive of European Integration (digital collection); PhilSci-Archive (preprints)

Percentage of journals that are peer reviewed: 95

Campus partners: campus departments or programs; individual faculty; graduate students; undergraduate students

Other partners: American Forensic Association; Brunel University; Consortium of Indonesian Universities–Pittsburgh (KPTIP); Fonds Ricoeur; Grupo Biblios: International Network of the Development of Library and Information Science; Institute for Linguistic Evidence; Institute of Integrative Omics and Applied Biotechnology; Institute of Public Health, Bangalore, India; Instituto Internacional de Literatura Iberoamericana; Kadir Has University; LAPS/ENSP

HIGHLIGHTED PUBLICATION

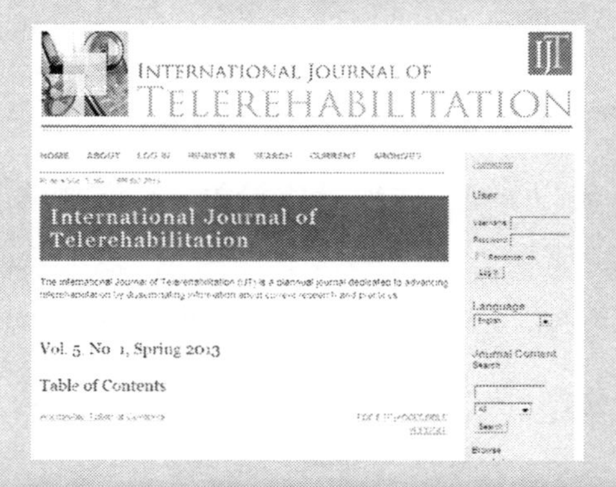

The International Journal of Telerehabilitation (IJT) is a biannual journal dedicated to advancing telerehabilitation by disseminating information about current research and practices.

lawreview.law.pitt.edu

OSwaldo Cruz Foundation LAPS; Motivational Interviewing Network of Trainers (MINT); Pennsylvania Library Association; Société Américaine de Philosophie de Langue Française; Society for Ricoeur Studies; TALE: The Association for Linguistic Evidence; University of Chapeco, Department of Anthropology; University of Kingston Centre for Modern European Philosophy

Publishing platform(s): Eprints; Fedora; Islandora; WordPress; locally developed software

Digital preservation strategy: discoverygarden; HathiTrust; LOCKSS; in-house

Additional services: graphic design (print or web); marketing; outreach; training; analytics; cataloging; metadata; compiling indexes and/or TOCs; notification of A&I sources; ISSN registration; DOI assignment/allocation of identifiers; dataset management; business model development; contract/license preparation; author copyright advisory; other author advisory; digitization; hosting of supplemental content; audio/video streaming

UNIVERSITY OF SAN DIEGO

Copley Library

Primary Unit: Special Collections and Archives

Primary Contact:
Kelly Riddle
Digital Initiatives Librarian
619-260-6850
kriddle@sandiego.edu

PROGRAM OVERVIEW

Mission/description: Digital publishing at the University of San Diego's Copley Library offers the university community the opportunity to share research, scholarly works, and other unique resources of historical or intellectual value. The library's digital publishing program will serve to advance faculty and student success and will foster intellectual collaboration both locally and globally. The library is dedicated to developing publishing services that will support and disseminate knowledge created or sponsored by the university so that it is readily discoverable, openly accessible, preserved, and sustainable. A goal of digital publishing will be to introduce faculty to a variety of new publishing models.

Year publishing activities began: 2013

Organization: centralized library publishing unit/department

Staff in support of publishing activities (FTE): library staff (1.5); undergraduate students (2)

Funding sources (%): library operating budget (100)

PUBLISHING ACTIVITIES

Media formats: text; images; audio; video; data; concept maps/modeling maps/visualizations; multimedia/interactive content

Campus partners: campus departments or programs; individual faculty; graduate students; undergraduate students

Publishing platform(s): bepress (Digital Commons); ContentPro

Digital preservation strategy: digital preservation services under discussion

Additional services: outreach; training; cataloging; metadata; open URL support; author copyright advisory; other author advisory; digitization; hosting of supplemental content; audio/video streaming

UNIVERSITY OF SOUTH FLORIDA

Tampa Library

Primary Unit: Academic resources
scholarcommons@usf.edu

Primary Contact:
Rebel Cummings-Sauls
Library Operations Coordinator
813-974-7381
rebelcs@usf.edu

Website: scholarcommons.usf.edu

PROGRAM OVERVIEW

Mission/description: The USF Tampa Library strives to develop and encourage research collaboration and initiatives throughout all areas of campus. Members of the USF community are encouraged to deposit their research with Scholar Commons. We commit to assisting faculty, staff, and students in all stages of the deposit process, to managing their work to optimize access/readership, and to ensure long-term preservation. Long-term preservation and increasing accessibility will increase citation rates and highlight the research accomplishments of this campus. Scholar Commons will have a direct impact on the University's four strategic goals: student success, research innovation, sound financial management, and creating new partnerships.

Year publishing activities began: 2007

Organization: centralized library publishing unit/department

Staff in support of publishing activities (FTE): library staff (3); graduate students (0.25)

Funding sources (%): library operating budget (60); endowment income (40)

PUBLISHING ACTIVITIES

Types of publications: journals produced under contract/MOU for external groups (12); monographs (2); textbooks; (4); technical/research reports (75); faculty conference papers and proceedings (350); student conference papers and proceedings (285); newsletters (10); databases (5); ETDs (4452); undergraduate capstone/honors theses (90); oral histories; events and lectures; course material; grey/white works

Media formats: text; images; audio; video; data; concept maps/modeling maps/ visualizations; multimedia/interactive content

Disciplinary specialties: geology and karst; Holocaust and genocide; environmental sustainability; literature; math/quantitative literature

Top publications: ETDs; *Social Science Research: Principle, Methods, and Practices* (journal); *International Journal of Speleology* (journal); *Journal of Strategic Security* (journal); *Studia UBB Geologia* (journal)

Percentage of journals that are peer reviewed: 92

Campus partners: campus departments or programs; individual faculty; graduate students

Other partners: National Cave and Karst Research Institute (NCKRI); Aphra Behn Society; Union Internationale de Spéléologie; Center for Conflict Management (CCM) of the National University of Rwanda (NUR); Henley-Putnam University; National Numeracy Network (NNN); IAVCEI Commission on Statistics in Volcanology (COSIV); Babeş-Bolyai University; National Center for Suburban Studies at Hofstra University

Publishing platform(s): bepress (Digital Commons)

Digital preservation strategy: LOCKSS; Portico; in-house; digital preservation services under discussion. PLN is being discussed. Bepress also offers preservation and backups.

Additional services: graphic design (print or web); typesetting; marketing; outreach; training; analytics; cataloging; metadata; compiling indexes and/or TOCs; notification of A&I sources; ISSN registration; DOI assignment/allocation of identifiers; open URL support; dataset management; peer review management; author copyright advisory; digitization; hosting of supplemental content; audio/ video streaming; add DOIs to references; suggest POD services

Plans for expansion/future directions: Adding a Coordinator role; expanding all content areas; and we currently have three new journals in process.

UNIVERSITY OF TENNESSEE

University of Tennessee Libraries

Founding Institution — Library Publishing Coalition

Primary Unit: Digital Production and Publishing/Newfound Press

Primary Contact:
Holly Mercer
Associate Dean for Scholarly Communication & Research Services
865-974-6899
hollymercer@utk.edu

Website: www.newfoundpress.utk.edu; trace.tennessee.edu

PROGRAM OVERVIEW

Mission/description: The University of Tennessee Libraries has developed a framework to make scholarly and specialized works available worldwide. Newfound Press, the University Libraries digital imprint, advances the community of learning by experimenting with effective and open systems of scholarly communication. Drawing on the resources that the university has invested in digital library development, Newfound Press collaborates with authors and researchers to bring new forms of publication to an expanding scholarly universe. UT Libraries provides open access publishing services, copyright education, and services to help scholars meet new data management and sharing requirements. In addition, we create digital collections of regional and global importance to support research and teaching.

Year publishing activities began: 2005

Organization: centralized library publishing unit/department

Staff in support of publishing activities (FTE): library staff (1.35); graduate students (0.5)

Funding sources (%): library operating budget (100)

PUBLISHING ACTIVITIES

Types of publications: faculty-driven journals (2); student-driven journals (1); journals produced under contract/MOU for external groups (1); monographs (3); faculty conference papers and proceedings (50); student conference papers and proceedings (23); newsletters (2); databases (1); ETDs (652); undergraduate capstone/honors theses (75)

Media formats: text; images; audio; video; data; multimedia/interactive content

Disciplinary specialties: East Tennessee; Great Smoky Mountains; anthropology; sociology; law

Top publications: *The Fishes of Tennessee* (monograph); *Building Bridges in Anthropology* (monograph); *To Advance Their Opportunities: Federal Policies Toward African American Workers from World War I to the Civil Rights Act of 1964* (monograph); *Goodness Gracious, Miss Agnes: Patchwork of Country Living* (monograph); "Why We Don't Vote: Low Voter Turnout in U.S. Presidential Elections" (thesis)

Percentage of journals that are peer reviewed: 50

Campus partners: UT Press; campus departments or programs; individual faculty; graduate students; undergraduate students

Other partners: Southern Anthropological Society; Music Theory Society of the Mid-Atlantic

Publishing platform(s): bepress (Digital Commons); locally developed software

Digital preservation strategy: DuraCloud/DSpace; MetaArchive

Additional services: graphic design (print or web); typesetting; copy-editing; marketing; analytics; cataloging; metadata; DOI assignment/allocation of identifiers; dataset management; peer review management; author copyright advisory; digitization; hosting of supplemental content; audio/video streaming; assignment of ISBNs

Plans for expansion/future directions: Exploring how to cultivate data publishing and how to support digital humanities on campus.

HIGHLIGHTED PUBLICATION

The Wondrous Bird's Nest I & II (*Das wunderbarliche Vogelnest*) is the only complete English translation of the fourth of the five Simplican novels by seventeenth-century German-language novelist Grimmelshausen.

newfoundpress.utk.edu/pubs/hiller

UNIVERSITY OF TEXAS AT SAN ANTONIO

University of Texas at San Antonio Libraries

Primary Unit: Learning Technology

Primary Contact:
Posie Aagaard
Assistant Dean for Collections and Curriculum Support
210-458-4878
posie.aagaard@utsa.edu

PROGRAM OVERVIEW

Mission/description: The UTSA Libraries collaborate with faculty to disseminate original scholarly content using a variety of platforms, ensuring open access while simultaneously acknowledging reader preferences.

Year publishing activities began: 2012

Organization: services are distributed across library units/departments

Staff in support of publishing activities (FTE): library staff (0.1)

Funding sources (%): library operating budget (100)

PUBLISHING ACTIVITIES

Types of publications: faculty conference papers and proceedings (1)

Media formats: text; images; video; concept maps/modeling maps/visualizations

Disciplinary specialties: astronomy

Top publications: *Torus Workshop 2012* (conference proceedings)

Campus partners: individual faculty; graduate students

Other partners: Science Organizing Committee

Publishing platform(s): CONTENTdm; WorldCat.org

Digital preservation strategy: in-house. Master copy is retained in a preferred file format; copies of the files are kept on local server (which has security, disaster recovery, and backup features) and also with OCLC; metadata has been created to support ongoing longevity.

Additional services: graphic design (print or web); typesetting; copy-editing; marketing; outreach; training; analytics; cataloging; metadata; contract/license preparation; author copyright advisory; hosting of supplemental content; audio/video streaming

Additional information: For our pilot publishing project, we collaborated with faculty who expressed a strong preference for using iBooks/iTunes as a publishing platform because the primary audience for the material (astronomy scholars) prefer to consume content on iPads. In addition to producing an iBook, we produced a multimedia-PDF, converting the content to a more open format for wider access and preservation purposes.

Plans for expansion/future directions: Actively seeking new opportunities to collaborate with faculty on publishing projects.

UNIVERSITY OF TORONTO

University of Toronto Libraries

Primary Unit: Information Technology Services

Primary Contact:
Sian Meikle
Interim Director, ITS
416-946-3689
sian.meikle@utoronto.ca

Website: jps.library.utoronto.ca; tspace.library.utoronto.ca

PROGRAM OVERVIEW

Mission/description: The University of Toronto Libraries maintains both the Open Journal System (OJS) and T-Space, the university's research repository with the aim to preserve and make available the university's scholarly contributions. We provide leadership and actively support scholarly communication needs by developing alternative forms of publication and viability models for the future that ensure the production and capture of research output.

Year publishing activities began: 2003

Organization: services are distributed across several campuses

Staff in support of publishing activities (FTE): library staff (3.75); graduate students (2)

Funding sources (%): library operating budget (100)

PUBLISHING ACTIVITIES

Types of publications: faculty-driven journals (25); student-driven journals (15); faculty conference papers and proceedings (170); student conference papers and proceedings (30); ETDs (6000); undergraduate capstone/honors theses (30)

Media formats: text; images; audio; video; data; concept maps/modeling maps/ visualizations; multimedia/interactive content

Disciplinary specialties: medicine/health sciences; humanities; social sciences; physical/natural sciences

Percentage of journals that are peer reviewed: 75

Campus partners: campus departments or programs; individual faculty; graduate students; undergraduate students

Other partners: University of Toronto Press; Ontario Council of University Libraries (OCUL); Canadian Association of Research Libraries (CARL)

Publishing platform(s): CONTENTdm; DSpace; Fedora; Islandora; OJS/OCS/OMP; WordPress; bibapp

Digital preservation strategy: Archive-It; DuraCloud/DSpace; LOCKSS; Scholars Portal; Synergies; Internet Archive

Additional services: graphic design (print or web); outreach; training; cataloging; metadata; business model development; contract/license preparation; author copyright advisory; digitization; audio/video streaming

Plans for expansion/future directions: Aligning focus.library.utoronto.ca (a more outwardly facing system for faculty profiling) with T-Space, the repository; working on copyright issues with our recently hired Scholarly Communication/Copyright Librarian.

UNIVERSITY OF UTAH

J. Willard Marriott Library

Primary Unit: Information Technology

Primary Contact:
John Herbert
Head, Digital Ventures
801-585-6019
john.herbert@utah.edu

PROGRAM OVERVIEW

Year publishing activities began: 2010

Organization: services are distributed across library units/departments

Staff in support of publishing activities (FTE): library staff (1)

Funding sources (%): library operating budget (100)

PUBLISHING ACTIVITIES

Types of publications: faculty-driven journals (3); student-driven journals (6); ETDs (100)

Media formats: text; images; audio; video; concept maps/modeling maps/visualizations

Disciplinary specialties: law; environmental studies; foreign languages; political science

Top publications: *Utah Law Review* (journal); *Hinckley Journal of Politics* (journal); *Utah Foreign Language Review* (journal); *Utah Environmental Law Review* (journal)

Percentage of journals that are peer reviewed: 75

Campus partners: individual faculty; graduate students; undergraduate students

Publishing platform(s): CONTENTdm; OJS/OCS/OMP; WordPress

Digital preservation strategy: Rosetta

Additional services: graphic design (print or web); outreach; metadata; DOI assignment/allocation of identifiers; author copyright advisory; digitization; hosting of supplemental content; audio/video streaming

HIGHLIGHTED PUBLICATION

The *Utah Historical Review* is the journal of student history published by the Alpha Rho chapter of Phi Alpha Theta (national history honor society) at the University of Utah.

utahhistoricalreview.com

UNIVERSITY OF VICTORIA

University of Victoria Libraries

Primary Unit: Scholarly Publishing Office
press@uvic.ca

Primary Contact:
Inba Kehoe
Scholarly Communications Librarian
250-472-5017
press@uvic.ca

Website: journals@uvic.ca; dspace.library.uvic.ca:8443

PROGRAM OVERVIEW

Mission/description: UVic Press represents the scholarly publishing expertise for the University of Victoria and its partner institutions and associations. We are dedicated to the online dissemination of knowledge and research through open access of journals, monographs, and other forms of publication. UVic Press offers an imprint to scholarship of a high quality, determined through peer review. We will work with emerging writers and research to promote success in scholarly publishing.

Year publishing activities began: 2008

Organization: services are distributed across library units/departments

Staff in support of publishing activities (FTE): library staff (1)

PUBLISHING ACTIVITIES

Types of publications: faculty-driven journals (2); student-driven journals (4); journals produced under contract/MOU for external groups (1); monographs (6); ETDs (398)

Media formats: text; images

Disciplinary specialties: humanities; social sciences; disability services; writing; creative fiction

Top publications: *Philosophy in Review* (journal); *Working Papers of the Linguistics Circle* (journal); *International Journal of Child, Youth and Family Studies* (journal); *Canadian Zooarchaeology* (journal); *Appeal: Review of Current Law and Law Reform* (journal)

Percentage of journals that are peer reviewed: 18

Campus partners: campus departments or programs; individual faculty

Other partners: Public Knowledge Project; Canadian Associate of Learned Journals; Universities Art Association of Canada; Association for Borderlands Studies

Publishing platform(s): DSpace; OJS/OCS/OMP

Digital preservation strategy: COPPUL; LOCKSS; Synergies

Additional services: copy-editing; marketing; outreach; training; analytics; cataloging; metadata; compiling indexes and/or TOCs; ISSN registration; contract/license preparation; author copyright advisory; digitization; hosting of supplemental content

Plans for expansion/future directions: Developing a fully functional university publishing program that will include publishing of journals, conference proceedings, and books. The program will include various imprints under the university press umbrella.

UNIVERSITY OF WASHINGTON

University of Washington Libraries

Primary Unit: Digital Initiatives

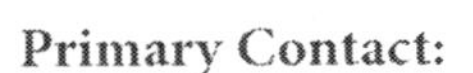

Primary Contact:
Ann Lally
Head, Digital Initiatives
206-685-1473
alally@uw.edu

Website: researchworks.lib.washington.edu

PROGRAM OVERVIEW

Mission/description: The University of Washington Libraries ResearchWorks Service provides faculty, researchers, and students with tools to archive and/or publish the products of research including datasets, monographs, images, journal articles, and technical reports.

Year publishing activities began: 1998

Organization: services are distributed across several campuses

Staff in support of publishing activities (FTE): library staff (1.5); graduate students (0.25)

Funding sources (%): library operating budget (100)

PUBLISHING ACTIVITIES

Types of publications: faculty-driven journals (2); journals produced under contract/MOU for external groups (2); technical/research reports (9); faculty conference papers and proceedings (1); ETDs (1300); undergraduate capstone/honors theses (30); research notebooks

Media formats: text; images; audio; video; data; concept maps/modeling maps/visualizations

Disciplinary specialties: information studies; anthropology; fisheries; Native American studies

Percentage of journals that are peer reviewed: 0

Campus partners: campus departments or programs; individual faculty; graduate students; undergraduate students

Other partners: Indo-Pacific Prehistory Association; Society for Slovene Studies

Publishing platform(s): CONTENTdm; DSpace; OJS/OCS/OMP

Digital preservation strategy: University eScience dark archive

Additional services: graphic design (print or web); training; analytics; cataloging; metadata; ISSN registration; DOI assignment/allocation of identifiers; peer review management; contract/license preparation; author copyright advisory; digitization; hosting of supplemental content

UNIVERSITY OF WATERLOO

University of Waterloo Library

Primary Unit: Digital Initiatives

Primary Contact:
Pascal Calarco
AUL, Research & Digital Discovery Services
519-888-4567 ext. 38215
pvcalarco@uwaterloo.ca

PROGRAM OVERVIEW

Mission/description: Enabling original scholarly research at the University of Waterloo from faculty, students, and staff.

Year publishing activities began: 1998

Organization: services are distributed across campus

Staff in support of publishing activities (FTE): library staff (0.5)

Funding sources (%): library operating budget (100)

PUBLISHING ACTIVITIES

Types of publications: student-driven journals (1); journals produced under contract/MOU for external groups (3); newsletters (6); databases (1); ETDs (840)

Media formats: text; images; audio; video; multimedia/interactive content

Disciplinary specialties: disability studies; mechanical engineering; sociology and criminology; food science

Top publications: ENGINE: Pre-Print Server for IEEE Society for Vehicular Technology (preprints); *Canadian Journal of Disability Studies* (journal); *Canadian Graduate Journal of Sociology and Criminology* (journal); *Canadian Journal of Food Safety* (journal)

Percentage of journals that are peer reviewed: 100

Campus partners: campus departments or programs; individual faculty; graduate students

Other partners: Theses Canada; Canadian Disability Studies Association; Canadian Association of Food Safety

Publishing platform(s): DSpace; OJS/OCS/OMP; locally developed software

Digital preservation strategy: Archive-It; Scholars Portal; in-house; digital preservation services under discussion; Theses Canada

Additional services: analytics; cataloging; metadata; ISSN registration; business model development; author copyright advisory; digitization; hosting of supplemental content

Additional information: We have also participated in the Networked Digital Library of Theses and Dissertations since 1998.

Plans for expansion/future directions: Extending to working papers, pre-prints, senior undergraduate work, and other original efforts.

UNIVERSITY OF WINDSOR

Leddy Library

Primary Unit: Information Services

Primary Contact:
Dave Johnston
Information Services Librarian, Scholarly Communications Coordinator
519-253-3000 ext. 3208
djohnst@uwindsor.ca

Website: scholar.uwindsor.ca; ojs.uwindsor.ca/ojs/leddy/index.php; ocs.uwindsor.ca/ocs/index.php/PC/virtues

Social media: facebook.com/Leddy.Library

PROGRAM OVERVIEW

Mission/description: The Leddy Library supports the dissemination of new scholarship by graduate, faculty, and staff researchers at the University of Windsor in a variety of forms. Through the Scholarship at UWindsor repository, we are able to support the dissemination of theses and dissertations and thus provide increased visibility to the work of our graduate students. We also use the repository to support conferences run on our campus by helping the organizers manage the submission workflow and publication process. As a longstanding supporter of Open Journal Systems, the library helps to publish and maintain several journals run from our campus, and we are currently in the process of using the new Open Monograph Press software to help support electronic monograph publishing. Providing support for open access is a central concern in all of our publishing endeavors. We seek to educate our users about the value of open access and to encourage various forms of open access publication.

Organization: services are distributed across library units/departments

Staff in support of publishing activities (FTE): library staff (2)

Funding sources (%): library operating budget (100)

PUBLISHING ACTIVITIES

Types of publications: faculty-driven journals (6); faculty conference papers and proceedings (120); ETDs (109)

Media formats: text; images

Disciplinary specialties: philosophy (information logic); social justice; scholarship of teaching and learning; philosophy (phenomenology); multivariate statistical techniques

Top publications: *Informal Logic* (journal); *Collected Essays in Teaching and Learning* (journal); *Applied Multivariate Research* (journal); *Studies in Social Justice* (journal); *PhaenEx* (journal)

Percentage of journals that are peer reviewed: 100

Campus partners: campus departments or programs; individual faculty; graduate students

Publishing platform(s): bepress (Digital Commons); OJS/OCS/OMP

Digital preservation strategy: LOCKSS

Additional services: marketing; training; analytics; cataloging; metadata; ISSN registration; DOI assignment/allocation of identifiers; open URL support; author copyright advisory; digitization; hosting of supplemental content

Plans for expansion/future directions: Extending use of existing systems to support the publication of more journals and conferences; launching an open monograph series with the philosophy department.

UNIVERSITY OF WISCONSIN–MADISON

University of Wisconsin–Madison Libraries

Primary Unit: General Library System

Primary Contact:
Elisabeth Owens
Special Assistant to the Vice Provost for Libraries
608-262-2566
eowens@library.wisc.edu

Website: parallelpress.library.wisc.edu; uwdc.library.wisc.edu

PROGRAM OVERVIEW

Mission/description: The General Library System publishes print and digital works featuring new works of scholars, researchers, and poets, and important scholarly and historical materials that are available for study in both print and digital formats. These publications are the result of collaborations with the scholarly community and represent an ongoing commitment by the Libraries to scholarly communication as a contribution to the Wisconsin Idea and in support of the outreach mission of the university.

Year publishing activities began: 1999

Organization: services are distributed across library units/departments

Staff in support of publishing activities (FTE): library staff (1); graduate students (0.5); undergraduate students (0.5)

Funding sources (%): library operating budget (70); non-library campus budget (10); endowment income (5); charitable contributions/Friends of the Library organizations (5); sales revenue (5); other (5)

PUBLISHING ACTIVITIES

Types of publications: faculty-driven journals (3); student-driven journals (1); monographs (10); faculty conference papers and proceedings (25); newsletters (1); ETDs (500); reformatted works

Media formats: text; images; audio; video; data; concept maps/modeling maps/ visualizations

Disciplinary specialties: University of Wisconsin; state of Wisconsin; African studies; ecology and natural resources; decorative arts and material culture

Top publications: WI Land Survey Records (digital collection); Foreign Relations of the United States (digital collection); Icelandic Online (digital collection); Africa Focus (digital collection); Decorative Arts Library (digital collection)

Percentage of journals that are peer reviewed: 75

Campus partners: campus departments or programs; individual faculty

Publishing platform(s): DSpace; Fedora; OJS/OCS/OMP; WordPress; locally developed software

Digital preservation strategy: CLOCKSS; HathiTrust; LOCKSS; in-house; digital preservation services under discussion

Additional services: graphic design (print or web); copy-editing; marketing; outreach; training; analytics; cataloging; metadata; ISSN registration; DOI assignment/allocation of identifiers; open URL support; dataset management; peer review management; budget preparation; contract/license preparation; author copyright advisory; other author advisory; digitization; hosting of supplemental content; audio/video streaming

Additional information: The majority of the Libraries' publishing activities involve the reformatting and dissemination of new versions of existing resources. We do publish new material, and our responses are primarily reflective of these activities (as opposed to our digital collections and repository services).

Plans for expansion/future directions: Increasing emphasis on open access publications and unique archival and special collections materials.

UTAH STATE UNIVERSITY

Merrill-Cazier Library

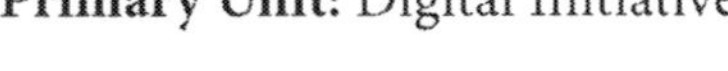

Primary Unit: Digital Initiatives

Primary Contact:
Becky Thoms
Copyright Librarian
435-797-0816
becky.thoms@usu.edu

Website: digitalcommons.usu.edu

PROGRAM OVERVIEW

Mission/description: USU Libraries is committed to the open dissemination of knowledge, as well as its delivery in new forms. Our publishing efforts emphasize open access and a commitment to look beyond traditional monographs and scholarly articles to disseminate dynamic scholarly works that can incorporate multimedia and social communications-style input.

Year publishing activities began: 2009

Organization: centralized library publishing unit/department

Staff in support of publishing activities (FTE): library staff (0.15); undergraduate students (0.5)

Funding sources (%): library operating budget (100)

PUBLISHING ACTIVITIES

Types of publications: faculty-driven journals (3); student-driven journals (1); journals produced under contract/MOU for external groups (1); monographs (3); faculty conference papers and proceedings (1811); newsletters (2); ETDs (311); faculty and student posters

Media formats: text; images; audio; video; data

Top publications: *Journal of Indigenous Research* (journal); *Journal of Mormon History* (journal); *Journal of Western Archives* (journal); *Foundations of Wave Phenomena* (journal); *An Introduction to Editing Manuscripts for Medievalists* (monograph)

Percentage of journals that are peer reviewed: 70

Campus partners: campus departments or programs; individual faculty; graduate students

Publishing platform(s): bepress (Digital Commons)

Digital preservation strategy: in-house. Our digital publishing content is archived on bepress servers in several geographic locations; we also archive copies on an in-house server. Some titles are preserved in HathiTrust, and we are investigating DPN.

Additional services: graphic design (print or web); cataloging; metadata; author copyright advisory; digitization

Plans for expansion/future directions: Building on existing collaborative relationship with the USU Press to connect authors with freelance providers of traditional publisher services such as peer review management, copy-editing, and typesetting.

HIGHLIGHTED PUBLICATION

Folklore and the Internet is a pioneering examination of the folkloric qualities of the World Wide Web, e-mail, and related digital media. It shows that folk culture, sustained by a new and evolving vernacular, has been a key to language, practice, and interaction online.

digitalcommons.usu.edu/usupress_pubs/35

VALPARAISO UNIVERSITY

Christopher Center for Library and Information Resources

Library Publishing Coalition
Contributing Institution

Primary Unit: Christopher Center Library Services
scholar@valpo.edu

Primary Contact:
Jonathan Bull
Scholarly Communication Services Librarian
219-464-5771
jon.bull@valpo.edu

Website: scholar.valpo.edu

PROGRAM OVERVIEW

Mission/description: ValpoScholar, a service of the Christopher Center Library and the Valparaiso University Law Library, is a digital repository and publication platform designed to collect, preserve, and make accessible the academic output of Valpo faculty, students, staff, and affiliates.

Year publishing activities began: 2011

Organization: services are distributed across two libraries, the Christopher Center and the Law Library

Staff in support of publishing activities (FTE): library staff (2); graduate students (1); undergraduate students (2)

PUBLISHING ACTIVITIES

Types of publications: faculty-driven journals (4); student-driven journals (3); technical/research reports (5); faculty conference papers and proceedings (10); student conference papers and proceedings (125); newsletters (1); ETDs (20); undergraduate capstone/honors theses (30); other conference proceedings

Media formats: text; images; audio; video; data

Disciplinary specialties: business and leadership ethics; creative writing (fiction); law

Top publications: *Valparaiso Law Review* (journal); *Valparaiso Fiction Review* (journal); *The Journal of Values-Based Leadership* (journal); *Third World Legal Studies* (journal)

Percentage of journals that are peer reviewed: 15

Campus partners: campus departments or programs; individual faculty; undergraduate students

Publishing platform(s): bepress (Digital Commons); CONTENTdm

Digital preservation strategy: no digital preservation services provided

Additional services: typesetting; marketing; outreach; training; analytics; metadata; ISSN registration; open URL support; dataset management; peer review management; author copyright advisory; other author advisory; digitization; hosting of supplemental content; audio/video streaming

Additional information: This is a growing service that appears to be needed and well-received on campus. We expect only growth in the future, along with external partnerships and more faculty-student collaboration.

VANDERBILT UNIVERSITY

Jean and Alexander Heard Library

Primary Unit: Scholarly Communications

Primary Contact:
Clifford B. Anderson
Director, Scholarly Communications
615-322-6938
clifford.anderson@vanderbilt.edu

Website: library.vanderbilt.edu/scholarly

PROGRAM OVERVIEW

Mission/description: The Jean and Alexander Heard Library fosters emerging modes of open access publishing by providing scholarly, technical, and financial support for the digital dissemination of faculty, student, and staff publications. The library maintains several publishing initiatives through its scholarly communication program. Currently, it publishes four peer-reviewed, open access journals—*AmeriQuests*, *Homiletic*, *Vanderbilt e-Journal of Luso-Hispanic Studies*, and the *Vanderbilt Undergraduate Research Journal*—using Open Journal Systems software. It also hosts a database of electronic theses and dissertations in cooperation with the Graduate School. Additionally, the library distributes undergraduate capstone projects through its institutional repository.

Year publishing activities began: 2004

Organization: services are distributed across library units/departments

Staff in support of publishing activities (FTE): library staff (2)

Funding sources (%): library operating budget (100)

PUBLISHING ACTIVITIES

Types of publications: faculty-driven journals (3); student-driven journals (1); ETDs (363); undergraduate capstone/honors theses (53)

Media formats: text; images

Disciplinary specialties: American studies; homiletics; Luso-Hispanic studies

Top publications: *AmeriQuests* (journal); *Homiletic* (journal); *Vanderbilt e-Journal of Luso-Hispanic Studies* (journal); *Vanderbilt Undergraduate Research Journal* (journal)

Percentage of journals that are peer reviewed: 100

Campus partners: campus departments or programs; individual faculty; graduate students; undergraduate students

Other partners: Academy of Homiletics

Publishing platform(s): DSpace; OJS/OCS/OMP; ETD-db

Digital preservation strategy: in-house; LOCKSS-ETD

Additional services: outreach; training; cataloging; author copyright advisory

Plans for expansion/future directions: Strengthening support for the publication of scientific datasets as well as projects in the digital humanities.

VILLANOVA UNIVERSITY

Falvey Memorial Library

Contributing Institution
Library Publishing Coalition

Primary Unit: Falvey Memorial Library

Primary Contact:
Darren G. Poley
Interim Library Director
610-519-6371
darren.poley@villanova.edu

PROGRAM OVERVIEW

Mission/description: In support of Villanova University's academic mission, the library is committed to the creation and dissemination of scholarship; utilizing digital modes and exploring new media for scholarly communication; and whenever possible, fostering open and public access to the intellectual contributions it publishes.

Organization: services are distributed across library units/departments

Staff in support of publishing activities (FTE): library staff (1.5)

Funding sources (%): library operating budget (100)

PUBLISHING ACTIVITIES

Types of publications: faculty-driven journals (3); journals produced under contract/MOU for external groups (2); undergraduate capstone/honors theses (5)

Media formats: text; images

Disciplinary specialties: American Catholic studies; Catholic higher education; theater; humanities; liberal arts and sciences

Top publications: *Journal of Catholic Higher Education* (journal); *American Catholic Studies* (journal); *Expositions* (journal); *Praxis* (journal); *Concept* (journal)

Percentage of journals that are peer reviewed: 100

Campus partners: campus departments or programs; individual faculty

Other partners: American Catholic Historical Society; Association of Catholic Colleges and Universities

Publishing platform(s): OJS/OCS/OMP

Digital preservation strategy: in-house

Additional services: graphic design (print or web); digitization

VIRGINIA COMMONWEALTH UNIVERSITY

VCU Libraries

Primary Unit: Information Management and Processing

Primary Contact:
John Duke
Senior Associate University Librarian
804-827-3624
jkduke@vcu.edu

Website: digarchive.library.vcu.edu

PROGRAM OVERVIEW

Mission statement: VCU's digital press provides the tools, infrastructure, and support for unique digital scholarly expressions from the VCU community of faculty and students from all disciplines.

Year publishing activities began: 2003

Organization: services are distributed across library units/departments

Total FTE in support of publishing activities: library staff (0.25)

Funding sources (%): library operating budget (100)

PUBLISHING ACTIVITIES

Types of publications: monographs (1), student conference papers and proceedings (11), ETDs (491)

Media formats: text; images; audio; video; concept maps/modeling/maps/ visualizations; multimedia/interactive content

Disciplinary specialties: history

Top publications: *British Virginia* (monograph); "Information Technology Outsourcing in U.S. Hospital Systems" (thesis); "A Computational Biology Approach to the Analysis of Complex Physiology" (thesis); "The Effects of the Handwriting Without Tears Program" (thesis); "Psychology and the Theater" (thesis)

Internal partners: campus departments or programs; individual faculty; graduate students; undergraduate students

Publishing platform(s): CONTENTdm; DSpace

Digital preservation strategy: in-house; digital preservation services under discussion

Additional services: marketing; outreach; training; cataloging; metadata; digitization

Additional information: VCU Libraries recruited a new professional position to advance research data management in the first quarter of academic year 2013-14; it expects to launch a library publishing program and a full institutional repository later this year.

Plans for expansion/future directions: Expanding the institutional repository to become a full partner in the SHARE initiative; creating a publishing platform for existing journals published by VCU faculty and for new scholarly journals and output from the entire VCU community.

VIRGINIA TECH

University Libraries

Primary Unit: Center for Digital Research and Scholarship

Primary Contact:
Gail McMillan
Director, Center for Digital Research and Scholarship Services
540-231-9252
gailmac@vt.edu

Website: scholar.lib.vt.edu; ejournals.lib.vt.edu; vtechworks.lib.vt.edu

PROGRAM OVERVIEW

Mission/description: The Libraries support the Virginia Tech community's needs (e.g., conference, journal, and book publishing; rights management and open access consulting, etc.) through digital publishing services. Virginia Tech has been hosting, providing access to, and preserving ejournals since 1989, but we are new to supporting the full workflow from article submission to peer review, editing, and production. We launched OJS in December 2012.

Year publishing activities began: 1989

Organization: centralized library publishing unit/department

Staff in support of publishing activities (FTE): library staff (1.5)

Funding sources (%): library operating budget (100)

PUBLISHING ACTIVITIES

Types of publications: faculty-driven journals (9); student-driven journals (1); journals produced under contract/MOU for external groups (6); faculty conference papers and proceedings (1); ETDs (900); yearbooks; annual reports

Media formats: text; images; audio; video; data; concept maps/modeling maps/visualizations; multimedia/interactive content

Disciplinary specialties: technology education

Top publications: ETDs; *Journal of Technology Education* (journal); *ALAN Review* (journal); *Journal of Industrial Teacher Education* (journal); *Journal of Technology Studies* (journal)

Percentage of journals that are peer reviewed: 100

Campus partners: campus departments or programs; individual faculty; graduate students; undergraduate students

Other partners: While our ejournal editors largely work on behalf of scholarly societies, we do not work directly with the societies.

Publishing platform(s): DSpace; OJS/OCS/OMP; locally developed software

Digital preservation strategy: LOCKSS; MetaArchive

Additional services: analytics; cataloging; metadata; DOI assignment/allocation of identifiers; dataset management; contract/license preparation; author copyright advisory; hosting of supplemental content; audio/video streaming

Plans for expansion/future directions: Consulting with editors about using OJS through CDRS Services; inviting hosted ejournal editors to consider using OJS; launching OCS; and collaborating with our university community to consider other publishing services.

HIGHLIGHTED PUBLICATION

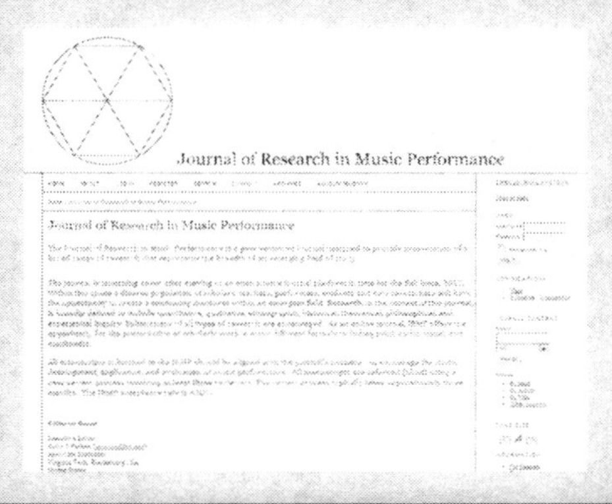

The *Journal of Research in Music Performance* is a peer-reviewed journal designed to provide presentation of a broad range of research that represents the breadth of an emerging field of study.

ejournals.lib.vt.edu/JRMP

WAKE FOREST UNIVERSITY

Z. Smith Reynolds Library

Primary Unit: Digital Publishing
kanewp@wfu.edu

Primary Contact:
William Kane
Digital Publishing
336-758-6181
kanewp@wfu.edu

Website: digitalpublishing.wfu.edu

PROGRAM OVERVIEW

Mission/description: Digital Publishing at Wake Forest University helps faculty, staff, and students create, collect, and convert previously or otherwise unpublished works into digitally distributed books, journals, articles, and the like.

Year publishing activities began: 2011

Organization: centralized library publishing unit/department

Staff in support of publishing activities (FTE): library staff (1)

Funding sources (%): library operating budget (50); non-library campus budget (50)

PUBLISHING ACTIVITIES

Types of publications: faculty-driven journals (1); student-driven journals (2); monographs (10); textbooks; (5); technical/research reports (5); faculty conference papers and proceedings (3); student conference papers and proceedings (3); newsletters (6); undergraduate capstone/honors theses (4)

Media formats: text; images; audio; video; data; concept maps/modeling maps/ visualizations; multimedia/interactive content

Percentage of journals that are peer reviewed: 0

Campus partners: campus departments or programs; individual faculty; graduate students; undergraduate students

Publishing platform(s): DSpace; Scalar; WordPress; Tizra

Digital preservation strategy: Amazon Glacier; Amazon S3; HathiTrust; in-house; digital preservation services under discussion

Additional services: graphic design (print or web); typesetting; copy-editing; marketing; outreach; training; analytics; cataloging; metadata; compiling indexes and/or TOCs; open URL support; business model development; budget preparation; contract/license preparation; author copyright advisory; digitization; audio/video streaming

Plans for expansion/future directions: Doubling the number of pages published year to year.

WASHINGTON UNIVERSITY IN ST. LOUIS

University Libraries

Library Publishing Coalition
FOUNDING INSTITUTION

Primary Unit: Digital Library Services
digital@wumail.wustl.edu

Primary Contact:
Emily Stenberg
Digital Publishing and Preservation Librarian
314-935-8329
emily.stenberg@wustl.edu

Website: openscholarship.wustl.edu

PROGRAM OVERVIEW

Mission/description: The mission of the Washington University in St. Louis Libraries Publishing Program is twofold: to provide alternatives to traditional publishing avenues, and to promote and disseminate original scholarly work of the Washington University community. Washington University Libraries began publishing ETDs in 2009, and in 2011, we launched the Open Scholarship repository to continue ETD publication, to provide a platform for the open access re-publication of faculty articles, and to provide for original publication of online journals and monographs. Since the launch of Open Scholarship, we have expanded into undergraduate honors theses and presentations, and have begun publishing monographs.

Year publishing activities began: 2009

Organization: centralized library publishing unit/department

Staff in support of publishing activities (FTE): library staff (1.5)

Funding sources (%): endowment income (100)

PUBLISHING ACTIVITIES

Types of publications: monographs (1); ETDs (264); undergraduate capstone/honors theses (13)

Media formats: text; images

Top publications: "Edith Wharton: Vision and Perception in Her Short Stories" (thesis); "Added-Tone Sonorities in the Choral Music of Eric Whitacre" (thesis); "Fashioning Women Under Totalitarian Regimes: 'New Women' of Nazi Germany and Soviet Russia" (thesis); "Computational Fluid Dynamics (CFD) Modeling of Mixed Convection Flows in Building Enclosures" (thesis); "Sentimental Ideology, Women's Pedagogy, and American Indian Women's Writing: 1815-1921" (thesis)

Campus partners: campus departments or programs; individual faculty; graduate students; undergraduate students

Publishing platform(s): bepress (Digital Commons)

Digital preservation strategy: in-house

Additional services: graphic design (print or web); copy-editing; metadata; DOI assignment/allocation of identifiers; other author advisory

Plans for expansion/future directions: Bringing a small number of journals (currently in development) online in the coming year.

WAYNE STATE UNIVERSITY

Wayne State University Library System

Primary Unit: Digital Publishing Unit

Primary Contact:
Joshua Neds-Fox
Coordinator for Digital Publishing
313-577-4460
jnf@wayne.edu

Program Overview
Mission/description: Wayne State's Digital Publishing Unit works to make unique, important, or institutionally relevant scholarly content available to the world at large, in the context of the WSU Library System's digital platforms.

Year publishing activities began: 2010

Organization: centralized library publishing unit/department

Staff in support of publishing activities (FTE): library staff (3); graduate students (0.5)

Funding sources (%): library operating budget (100)

PUBLISHING ACTIVITIES

Types of publications: faculty-driven journals (1); student-driven journals (1); journals produced under contract/MOU for external groups (10); ETDs (200); undergraduate capstone/honors theses (12)

Media formats: text

Percentage of journals that are peer reviewed: 100

Campus partners: Wayne State University Press; campus departments or programs; individual faculty; graduate students; undergraduate students

Publishing platform(s): bepress (Digital Commons); Fedora

Digital preservation strategy: in-house; digital preservation services under discussion

Additional services: graphic design (print or web); typesetting; copy-editing; marketing; outreach; training; analytics; cataloging; metadata; author copyright advisory; other author advisory; digitization; hosting of supplemental content

HIGHLIGHTED PUBLICATION

JMASM is an independent, peer-reviewed, open access journal providing a scholarly outlet for applied (non)parametric statisticians, data analysts, researchers, psychometricians, quantitative or qualitative evaluators, and methodologists.

digitalcommons.wayne.edu/jmasm

WESTERN UNIVERSITY

Western Libraries

Primary Unit: Library Information Resources Management
wlscholcomm@uwo.ca

Primary Contact:
Karen Marshall
Assistant University Librarian
519-661-2111 ext. 84850
karen.marshall@uwo.ca

Website: ir.lib.uwo.ca

PROGRAM OVERVIEW

Mission/description: Scholarship@Western is a multi-functional portal that collects, showcases, archives, and preserves a variety of materials created or sponsored by the University of Western Ontario community. It aims to facilitate knowledge sharing and broaden the international recognition of Western's academic excellence by providing open access to Western's intellectual output and professional achievements. It also serves as a platform to support Western's scholarly communication needs and provides an avenue for the compliance of research funding agencies' open access policies.

Year publishing activities began: 2010

Organization: centralized library publishing unit/department

Staff in support of publishing activities (FTE): library staff (3)

PUBLISHING ACTIVITIES

Types of publications: faculty-driven journals (10); student-driven journals (10); journals produced under contract/MOU for external groups (2); monographs (6); technical/research reports (20); faculty conference papers and proceedings (30); student conference papers and proceedings (2); ETDs (50)

Media formats: text; images; audio; video; data

Percentage of journals that are peer reviewed: 90

Campus partners: campus departments or programs; individual faculty; graduate students; undergraduate students

Other partners: scholarly societies; conferences

LIBRARIES OUTSIDE THE UNITED STATES AND CANADA

AUSTRALIAN NATIONAL UNIVERSITY

Australian National University Library

Primary Contact:
Lorena Kanellopoulos
Manager, ANU E Press
+61-2-6125-4536
lorena.kanellopoulos@anu.edu.au

Website: epress.anu.edu.au; digitalcollections.anu.edu.au; anulib.anu.edu.au

PROGRAM OVERVIEW

Mission/description: The Library aims to support *ANU by 2020*'s goals of excellence in research and education and the University's role as a national policy resource.

Year publishing activities began: 2001

Organization: centralized library publishing unit/department

Staff in support of publishing activities (FTE): library staff (5)

Funding sources (%): library operating budget (100)

PUBLISHING ACTIVITIES

Types of publications: student-driven journals (4); monographs (61); faculty conference papers and proceedings (200); ETDs (130)

Media formats: text; images; audio; video

Percentage of journals that are peer reviewed: 100

Campus partners: campus departments or programs; individual faculty

Publishing platform(s): DSpace; WordPress

Digital preservation strategy: no digital preservation services provided

Additional services: graphic design (print or web); cataloging; author copyright advisory

Plans for expansion/future directions: For key strategic directions, see anulib.anu.edu.au/_resources/reports-and-publications/publications/Library_operational_plan_draft_2013.pdf.

EDITH COWAN UNIVERSITY

Edith Cowan University Library

Primary Unit: Research Services
researchonline@ecu.edu.au

Primary Contact:
Julia Gross
Senior Librarian, Research Services
+61-8-6304-3698
j.gross@ecu.edu.au

Website: ro.ecu.edu.au

PROGRAM OVERVIEW

Year publishing activities began: 2010

Organization: centralized library publishing unit/department

Staff in support of publishing activities (FTE): library staff (3)

Funding sources (%): library materials budget (10); library operating budget (90)

PUBLISHING ACTIVITIES

Types of publications: faculty-driven journals (5); student-driven journals (1); faculty conference papers and proceedings (3); ETDs (97); undergraduate capstone/honors theses (39)

Media formats: text; images; concept maps/modeling maps/visualizations

Disciplinary specialties: education; business; social and behavioral sciences; medicine and health sciences; arts and humanities

Top publications: *Australian Journal of Teacher Education* (journal); *Landscapes* (journal); *eCulture* (journal); *Journal of Emergency Primary Health Care* (journal); *Research Journalism* (journal)

Percentage of journals that are peer reviewed: 100

Campus partners: campus departments or programs; individual faculty; undergraduate students

Publishing platform(s): bepress (Digital Commons)

Digital preservation strategy: digital preservation services under discussion

Additional services: marketing; outreach; training; analytics; cataloging; metadata; notification of A&I sources; ISSN registration; DOI assignment/allocation of identifiers; author copyright advisory; digitization

Plans for expansion/future directions: Increasing numbers of journals published; investigating ebook publication.

HUMBOLDT-UNIVERSITÄT ZU BERLIN

Universitätsbibliothek

Primary Unit: Arbeitsgruppe Elektronisches Publizieren

Primary Contact:
Niels Fromm
Head Electronic Publishing Group
+49-2093-70070
fromm@ub.hu-berlin.de

Website: edoc.hu-berlin.de

PROGRAM OVERVIEW

Mission/description: The edoc-Server is the Institutional Repository of Humboldt University. On this server every member of the University is able to publish his or her electronic theses and/or any documents as open access. We accept anything from single articles or volumes to series of open access publications.

Year publishing activities began: 1997

Organization: centralized library publishing unit/department

Staff in support of publishing activities (FTE): library staff (4); graduate students (1)

Funding sources (%): library operating budget (50); other (50)

PUBLISHING ACTIVITIES

Types of publications: faculty-driven journals (5); student-driven journals (1); monographs (200); technical/research reports (100); ETDs (300)

Media formats: text

Percentage of journals that are peer reviewed: 0

Campus partners: campus departments or programs

Publishing platform(s): locally developed software

Digital preservation strategy: CLOCKSS; LOCKSS; in-house

Additional services: cataloging; metadata; digitization; document templates for MS-Office; Styles for Endnote / Citavi

Plans for expansion/future directions: Developing a concept and a workflow for the publication of research data in addition to electronic theses.

MONASH UNIVERSITY

Monash University Library

Primary Unit: Research Infrastructure Division

Primary Contact:
Andrew Harrison
Research Repository Librarian
+61-3-9905-2682
andrew.harrison@monash.edu

PROGRAM OVERVIEW

Mission/description: Publishing at Monash University is carried out by Monash University Research Repository and Monash University Publishing, both of which are parts of the University Library. Monash University Research Repository is a digital archive of selected content representing Monash's research activity. The repository provides staff and students a place to deposit their research collections, data, or publications so they are centrally stored and managed, with the content easily discoverable online by their peers globally and by the broader community. The University requires that successful PhD theses are submitted to the repository for online publication. The repository is intended to be primarily an open access repository but does contain restricted access content on a case by case basis (e.g., embargoed theses). Monash University Publishing focuses on peer-reviewed monographs, which are published in both online open access and traditional print forms—as such it is not included here. It seeks to publish scholarly work of the highest quality, ensured by rigorous peer review; maximise the impact of those titles; represent the breadth and energy of Monash University research interests (while not excluding contributors from anywhere); promote the free exchange of knowledge; play a coordinating role in the production and dissemination of Monash's scholarly publications, and provide a body of publishing expertise within the University.

Year publishing activities began: 2000

Organization: centralized library publishing unit/department

Staff in support of publishing activities (FTE): library staff (3)

Funding sources (%): library operating budget (100)

PUBLISHING ACTIVITIES

Types of publications: faculty-driven journals (3); ETDs (450); research datasets such as images and sound files

Media formats: text; images; audio; data

Disciplinary specialties: Geographic Information Systems (GIS); comparative literature and cultural studies; social/community work

Top publications: *PAN: Philosophy Activism Nature* (journal); *Practice Reflexions* (journal); *Applied GIS* (journal)

Percentage of journals that are peer reviewed: 100

Campus partners: campus departments or programs; individual faculty

Other partners: Australian Community Workers Association

Publishing platform(s): Fedora; VITAL

Digital preservation strategy: in-house

Additional services: DOI assignment/allocation of identifiers

Additional information: Journal publishing is largely a legacy service. Our future focus is on theses and research data. We also think that separating the University Press from the repository is unhelpful: we see them as complimentary, and both are exploring new ways for Libraries to be involved in publishing going forward.

Plans for expansion/future directions: Expanding theses program to include master's and PhD candidates from disciplines previously exempt from the compulsory submission process; expanding the range of research data included in the repository. Changes to Australian funding council rules will increase the amount of open access journal material we hold.